Seed Borne Diseases: Ecofriendly Management

Seed Borne Diseases: Ecofriendly Management

Brindra Chauhan

RANDOM PUBLICATIONS
NEW DELHI (INDIA)

Seed Borne Diseases: Ecofriendly Management

ISBN 978-93-5111-594-6

Published in 2015 in India by

RANDOM PUBLICATIONS

4376-A/4B, Gali Murari Lal, Ansari Road
New Delhi-110 002
Phone : +9111-43580356, 011-23289044, 011-43142548
e-mail: sales@randompublications.com,
info@randompublications.com, randomexports@gmail.com

Reprinted 2018

Type Setting by : Friends Media, Delhi-110089
Digitally Printed at : Replika Press Pvt. Ltd.

Preface

Some plant diseases are spread through the actual seeds themselves. When customers buy our seeds, we do our best to ensure that these seeds are disease-free. Non-organic companies can spray their seeds with chemical fungicides to kill off the diseases, but organic companies such as ours need to utilize different tactics. Seedborne disease refers to the particular plant diseases that are transmitted by seed. In some cases the transmission on seed is insignificant compared to the population of disease organisms that exist in soil or on weed species. In other cases, the transmission on seed is the primary means by which a disease spreads. While we are cautious about any type of disease on seed, it is this latter set of diseases that we must be most vigilant in controlling. For the purposes of this article, we will call them seed-specific diseases.

Diseases of plants are caused primarily by three types of pathogens: bacteria, fungi, and viruses. Despite that fungi comprise the largest group of pathogens, the bulk of seed-specific diseases are caused by bacteria or viruses. This is due to the fact that bacteria and viruses are more adept at entering and then traveling through the veins of the plant, a phenomenon known as 'systemic infection,' and from the vascular system may make their way into the developing embryos of seeds. Fungi, in contrast, tend to be restricted to the outer layers of the plant, where they initiate infection by means of air-borne spores and then proceed to spread by attacking nearby cells of the outer layers. Fungi are much less likely to enter the vascular system of the plant, and thus infect seed mostly when they either 'crawl' all the way to seed on the outside of the plant, or else send out spores that land on the seed. In either case, the fungal spores are on the outside of the seed, in the layers of the seed coat. Spores on the seed coat are more prone to either dry up and die, or else to get sloughed off with the seed coat during seed germination, thereby failing to cause disease on the next generation of plants.

I would like to thank my team for standing beside me throughout my career and writing this book. My special thanks go to "Random Publications" who have published the book.

– Brindra Chauhan

Contents

Preface .... *v-vi*

1. Seeds Germination 1

Seed 1
Seed Structure 1
Germination 12
Factors Affecting Seed Germination 25
Dormancy 27

2. Bacterial Growth in Seed 30

Introduction 30
Intrinsic Factors 30
Extrinsic Factors 45
Storage/holding Conditions 51

3. Germination Testing 57

Germination Testing in the Nursery 72
Combining Purity and Germination Tests 73
Indirect Tests of Viability 74
Testing Moisture Content 80

4. Seed Saving 88

Introduction 88
Process of Seed Saving 89
Saving Tomato Seed 99

5. Seed Treatment 104

Technology 104
Application Methods 106
Advantages 109

6. Seed Germination 111

Introduction 111
Pollination and Fertilization 111

Angiosperm Seed Development 112
Angiosperm Fruit Development 115

7. **Seed Drying** **125**

Introduction 125
Requirement of Seeds to be Driedup 125
When should Seeds be Dried 125
Process 125

8. **Seed-Borne Disease** **140**

Introduction 140
Disease Management in Seed Crops 141
Cause 142
Disease Spread 145

9. **Seed Testing Management** **161**

Introduction 161
Sampling 163
Puriy Analsis 169
Seed Weight 171

10. **Effect of Preservation Technologies and Microbiological .... Inactivation** **174**

Introduction 174
Validation of Processing Parameters 175
Processing Technologies 175
Selection of Challenge Organisms 182

11. **Allosteric Enzymes** **194**

Introduction 194
Specificity 194
Controls Over Enzymatic Activity 195
Allosteric Regulation of Enzyme Activity 200

12. **Seed Habit, Phylogeny** **221**

Introduction 221
Existence 222
Inside of Seed 224

13. **Indian Agriculture and Biotechnology: The Challenge of the Next Millennium** **235**

Biotechnology in India: A Promising Future 235
Indian Advantage in Agriculture 236
India's Bio Industry: Segment Review 237

Bibliography **285**

Index **287**

1

Seeds Germination

In order to understand the true nature of a seed it is necessary to examine its origin and construction, and watch its development as far as possible from the earliest stages to the time when it gives rise to a completely formed young plant.

SEED

A seed is a small embryonic plant enclosed in a covering called the seed coat, usually with some stored food. It is the product of the ripened ovule of gymnosperm and angiosperm plants which occurs after fertilization and some growth within the mother plant. The formation of the seed completes the process of reproduction in seed plants (started with the development of flowers and pollination), with the embryo developed from the zygote and the seed coat from the integuments of the ovule. Seeds have been an important development in the reproduction and spread of flowering plants, relative to more primitive plants like mosses, ferns and liverworts, which do not have seeds and use other means to propagate themselves. This can be seen by the success of seed plants (both gymnosperms and angiosperms) in dominating biological niches on land, from forests to grasslands both in hot and cold climates. The term seed also has a general meaning that predates the above—anything that can be sown, *e.g.* "seed" potatoes, "seeds" of corn or sunflower "seeds". In the case of sunflower and corn "seeds", what is sown is the seed enclosed in a shell or hull, and the potato is a tuber.

SEED STRUCTURE

A typical seed includes three basic parts:

1. An embryo,
2. A supply of nutrients for the embryo, and
3. A seed coat.

The embryo is an immature plant from which a new plant will grow under proper conditions. The embryo has one cotyledon or seed leaf in monocotyledons, two cotyledons in almost all dicotyledons and two or more in gymnosperms. The radicle is the embryonic root. The plumule is the embryonic shoot. The embryonic stem above the point of attachment of the cotyledon(s) is the epicotyl. The embryonic stem below the point of attachment is the hypocotyl.

Within the seed, there usually is a store of nutrients for the seedling that will grow from the embryo. The form of the stored nutrition varies depending on the kind of plant. In angiosperms, the stored food begins as a tissue called the endosperm, which is derived from the parent plant via double fertilization. The usually triploid endosperm is rich in oil or starch and protein. In gymnosperms, such as conifers, the food storage tissue is part of the female gametophyte, a haploid tissue. In some species, the embryo is embedded in the endosperm or female gametophyte, which the seedling will use upon germination. In others, the endosperm is absorbed by the embryo as the latter grows within the developing seed, and the cotyledons of the embryo become filled with this stored food. At maturity, seeds of these species have no endosperm and are termed exalbuminous seeds. Some exalbuminous seeds are bean, pea, oak, walnut, squash, sunflower, and radish. Seeds with an endosperm at maturity are termed albuminous seeds. Most monocots (*e.g.* grasses and palms) and many dicots (*e.g.* brazil nut and castor bean) have albuminous seeds. All gymnosperm seeds are albuminous.

The seed coat (or testa) develops from the tissue, the integument, originally surrounding the ovule. The seed coat in the mature seed can be a paper–thin layer (*e.g.* peanut) or something more substantial (*e.g.* thick and hard in honey locust and coconut). The seed coat helps protect the embryo from mechanical injury and from drying out. In addition to the three basic seed parts, some seeds have an appendage on the seed coat such an aril (as in yew and nutmeg) or an elaiosome (as in Corydalis) or hairs (as in cotton). There may also be a scar on the seed coat, called the hilum; it is where the seed was attached to the ovary wall by the funiculus.

Kinds of Seeds

Many structures commonly referred to as "seeds" are actually dry fruits. Sunflower seeds are sold commercially while still enclosed within the hard wall of the fruit, which must be split open to reach the seed. Different groups of plants have other modifications, the so–called stone fruits (such as the peach) have a hardened fruit layer (the endocarp) fused to and surrounding the actual seed. Nuts are the one–seeded, hard shelled fruit, of some plants, with an indehiscent seed, such as an acorn or hazelnut.

Seed Production

Seeds are produced in several related groups of plants, and their manner of production distinguishes the angiosperms ("enclosed seeds") from the gymnosperms ("naked seeds"). Angiosperm seeds are produced in a hard or fleshy structure called a fruit that encloses the seeds, hence the name. (Some fruits have layers of both hard and fleshy material). In gymnosperms, no special structure develops to enclose the seeds, which begin their development "naked" on the bracts of cones. However, the seeds do become covered by the cone scales as they develop in some species of conifer.

Seed production in natural plant populations vary widely from year–to–year in response to weather variables, insects and diseases, and internal cycles within the plants themselves. Over a 20–year period, for example, forests composed of loblolly pine and shortleaf pine produced from 0 to nearly 5 million sound pine seeds per hectare. Over this period, there were six bumper seeds, five poor seeds crops, and nine good seed crops, when evaluated in regard to producing adequate seedlings for natural forest reproduction.

Seed Development

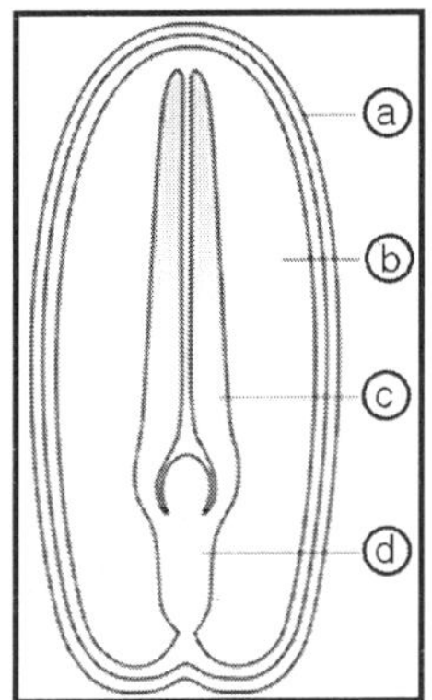

Fig. Diagram of the Internal Structure of a Dicot Seed and Embryo.

(a) Seed Coat,
(b) Endosperm,
(c) Cotyledon,
(d) Hypocotyl.

The seed, which is an embryo with two points of growth (one of which forms the stems the other the roots) is enclosed in a seed coat with some food reserves.

Angiosperm seeds consist of three genetically distinct constituents:

1. The embryo formed from the zygote,
2. The endosperm, which is normally triploid,
3. The seed coat from tissue derived from the maternal tissue of the ovule.

In angiosperms, the process of seed development begins with double fertilization and involves the fusion of the egg and sperm nuclei into a zygote. The second part of this process is the fusion of the polar nuclei with a second sperm cell nucleus, thus forming a primary endosperm. Right after fertilization, the zygote is mostly inactive but the primary endosperm divides rapidly to form the endosperm tissue.

This tissue becomes the food that the young plant will consume until the roots have developed after germination or it develops into a hard seed coat. The seed coat forms from the two integuments or outer layers of cells of the ovule, which derive from tissue from the mother plant, the inner integument forms the tegmen and the outer forms the testa.

When the seed coat forms from only one layer it is also called the testa, though not all such testa are homologous from one species to the next. In gymnosperms, the two sperm cells transferred from the pollen do not develop seed by double fertilization but one sperm nucleus unites with the egg nucleus and the other sperm is not used. Sometimes each sperm fertilizes an egg cell and one zygote is then aborted or absorbed during early development. The seed is composed of the embryo (the result of fertilization) and tissue from the mother plant, which also form a cone around the seed in coniferous plants like Pine and Spruce.

The ovules after fertilization develop into the seeds; the main parts of the ovule are the funicle; which attaches the ovule to the placenta, the nucellus; the main region of the ovule were the embryo sac develops, the micropyle; A small pore or opening in the ovule where the pollen tube usually enters during the process of fertilization, and the chalaza; the base of the ovule opposite the micropyle, where integument and nucellus are joined together.

The shape of the ovules as they develop often affects the finale shape of the seeds. Plants generally produce ovules of four shapes: the most common shape is called anatropous, with a curved shape. Orthotropous ovules are straight with all the parts of the ovule lined up in a long row producing an uncurved seed. Campylotropous ovules have a curved embryo sac often giving the seed a tight "c" shape. The last ovule shape is called amphitropous, where the ovule is partly inverted and turned back 90 degrees on its stalk or funicle.

In the majority of flowering plants, the zygote's first division is transversely oriented in regards to the long axis, and this establishes the polarity of the embryo. The upper or chalazal pole becomes the main area of growth of the embryo, while the lower or micropylar pole produces the stalk–like suspensor that attaches to the micropyle. The suspensor absorbs and manufacturers nutrients from the endosperm that are utilized during the embryos growth.

The embryo is composed of different parts; the epicotyle will grow into the shoot, the radicle grows into the primary root, the hypocotyl connects the epicotyle and the radicle, the cotyledons form the seed leaves, the testa or seed coat forms the outer covering of the seed. Monocotyledonous plants like corn, have other structures; instead of the hypocotyle–epicotyle, it has a coleoptile that forms the first leaf and connects to the coleorhiza that connects to the primary root and adventitious roots form from the sides. The seeds of corn are constructed with these structures; pericarp, scutellum (single large cotyledon) that absorbs nutrients from the endosperm, endosperm, plumule, radicle, coleoptile and coleorhiza—these last two structures are sheath–like and enclose the plumule and radicle, acting as a protective covering. The testa or seed coats of both monocots and dicots are often marked with patterns and textured markings, or have wings or tufts of hair.

Seed Size and Seed Set

Seeds are very diverse in size. The dust–like orchid seeds are the smallest with about one million seeds per gram; they are often embryonic seeds with immature embryos and no significant energy reserves. Orchids and a few other groups of plants are myco–heterotrophs which depend on mycorrhizal fungi for nutrition during germination and the early growth of the seedling. Some terrestrial Orchid seedlings, in fact, spend the first few years of their life deriving energy from the fungus and do not produce green leaves. At over 20 kg, the largest seed is the coco de mer.

Plants that produce smaller seeds can generate many more seeds per flower, while plants with larger seeds invest more resources into those seeds and normally produce fewer seeds. Small seeds are quicker to ripen and can be dispersed sooner, so fall blooming plants often have small seeds. Many annual plants produce great quantities of smaller seeds; this helps to ensure that at least a few will end in a favourable place for growth. Herbaceous perennials and woody plants often have larger seeds, they can produce seeds over many years, and larger seeds have more energy reserves for germination and seedling growth and produce larger, more established seedlings after germination.

Seed Functions

Seeds serve several functions for the plants that produce them. Key among these functions are nourishment of the embryo, dispersal to a new location, and dormancy during unfavourable conditions. Seeds fundamentally are a means of reproduction and most seeds are the product of sexual reproduction which produces a remixing of genetic material and phenotype variability that natural selection acts on.

Embryo Nourishment

Seeds protect and nourish the embryo or young plant. Seeds usually give a seedling a faster start than a sporeling from a spore, because of the larger food reserves in the seed and the multicellularity of the enclosed embryo.

SEED DISPERSAL

Unlike animals, plants are limited in their ability to seek out favourable conditions for life and growth. As a result, plants have evolved many ways to disperse their offspring by dispersing their seeds. A seed must somehow "arrive" at a location and be there at a time favourable for germination and growth. When the fruits open and release their seeds in a regular way, it is called dehiscent, which is often distinctive for related groups of plants, these fruits include; Capsules, follicles, legumes, silicles and siliques. When fruits do not open and release their seeds in a regular fashion they are called indehiscent, which include the fruits achenes, caryopsis, nuts, samaras, and utricles. Seed dispersal is seen most obviously in fruits; however many seeds aid in their own dispersal. Some kinds of seeds are dispersed while still inside a fruit or cone, which later opens or disintegrates to release the seeds. Other seeds are expelled or released from the fruit prior to dispersal. For example, milkweeds produce a fruit type, known as a follicle, that splits open along one side to release the seeds. Iris capsules split into three "valves" to release their seeds.

By wind (anemochory):

- Many seeds (*e.g.* maple, pine) have a wing that aids in wind dispersal.
- The dustlike seeds of orchids are carried efficiently by the wind.
- Some seeds, (*e.g.* dandelion, milkweed, poplar) have hairs that aid in wind dispersal.

Some winged seeds have two, and some have only one wing.

By water (hydrochory):

- Some plants, such as Mucuna and Dioclea, produce buoyant seeds termed sea–beans or drift seeds because they float in rivers to the oceans and wash up on beaches.

By animals (zoochory):

- Seeds (burrs) with barbs or hooks (*e.g.* acaena, burdock, dock) which attach to animal fur or feathers, and then drop off later.
- Seeds with a fleshy covering (*e.g.* apple, cherry, juniper) are eaten by animals (birds, mammals, reptiles, fish) which then disperse these seeds in their droppings.
- Seeds (nuts) which are an attractive long–term storable food resource for animals (*e.g.* acorns, hazelnut, walnut); the seeds are stored some distance from the parent plant, and some escape being eaten if the animal forgets them.

MYRMECOCHORY

Myrmecochory is the dispersal of seeds by ants. Foraging ants disperse seeds which have appendages called elaiosomes (*e.g.* bloodroot, trilliums, Acacias, and many species of Proteaceae). Elaiosomes are soft, fleshy structures that contain nutrients for animals that eat them. The ants carry such seeds back to their nest, where the elaiosomes are eaten. The remainder of the seed, which is hard and inedible to the ants, then germinates either within the nest or at a removal site where the seed has been discarded by the ants. This dispersal relationship is an example of mutualism, since the plants depend upon the ants to disperse seeds, while the ants depend upon the plants seeds for food. As a result, a drop in numbers of one partner can reduce success of the other. In South Africa, the Argentine ant (Linepithema humile) has invaded and displaced native species of ants. Unlike the native ant species, Argentine ants do not collect the seeds of Mimetes cucullatus or eat the elaiosomes. In areas where these ants have invaded, the numbers of Mimetes seedlings have dropped.

Seed Dormancy

Seed dormancy has two main functions: the first is synchronizing germination with the optimal conditions for survival of the resulting seedling; the second is spreading germination of a batch of seeds over time so that a catastrophe after germination (*e.g.* late frosts, drought, herbivory) does not result in the death of all offspring of a plant (bet–hedging). Seed dormancy is defined as a seed failing to germinate under environmental conditions optimal for germination, normally when the environment is at a suitable temperature with proper soil moisture. This true dormancy or innate dormancy is therefore caused by conditions within the seed that prevent germination. Thus dormancy is a state of the seed, not of the environment. Induced dormancy, enforced dormancy or seed quiescence occurs when a seed fails to germinate because the external environmental conditions are inappropriate for germination, mostly in response to conditions being too dark or light, too cold or hot, or too dry.

Seed dormancy is not the same as seed persistence in the soil or on the plant, though even in scientific publications dormancy and persistence are often confused or used as synonyms. Often seed dormancy is divided into four major categories: exogenous; endogenous; combinational; and secondary. A more recent system distinguishes five classes ofdormancy: morphological, physiological,morphophysiological, physical and combinational dormancy.

Exogenous dormancy is caused by conditions outside the embryo including:

- Physical dormancy or hard seed coats occurs when seeds are impermeable to water. At dormancy break a specialized structure, the 'water gap', is disrupted in response to environmental cues, especially temperature, so that water can enter the seed and germination can occur. Plant families where physical dormancy occurs include Anacardiaceae, Cannaceae, Convulvulaceae, Fabaceae and Malvaceae.
- Chemical dormancy considers species that lack physiological dormancy, but where a chemical prevents germination. This chemical can be leached out of the seed by rainwater or snow melt or be deactivated somehow. Leaching of chemical inhibitors from the seed

by rain water is often cited as an important cause of dormancy release in seeds of desert plants, however little evidence exists to support this claim.

Endogenous dormancy is caused by conditions within the embryo itself, including:

- Morphological dormancy where germination is prevented due to morphological characteristics of the embryo. In some species the embryo is just a mass of cells when seeds are dispersed, it is not differentiated. Before germination can take place both differentiation and growth of the embryo have to occur. In other species the embryo is differentiated but not fully grown (underdeveloped) at dispersal and embryo growth up to a species specific length is required before germination can occur. Examples of plant families where morphological dormancy occurs are Apiaceae, Cycadaceae, Liliaceae, Magnoliaceae and Ranunculaceae.
- Morphophysiological dormancy seeds with underdeveloped embryos, and which in addition have physiological components to dormancy. These seeds therefore require a dormancy–breaking treatments as well as a period of time to develop fully grownem bryos. Plant families where morphophysiological dormancy occurs include Apiaceae, Aquifoliaceae, Liliaceae, Magnoliaceae, Papaveraceae and Ranunculaceae. Some plants with morphophysi-ological dormancy like Asarum or Trillium species have multiple types of dormancy, one affects radicle (root) growth while the other affects plumule (shoot) growth. The terms "double dormancy" and "2–year seeds" are used for species whose seeds need two years to complete germination or at least two winters and one summer. Dormancy of the radicle (seedling root) is broken during the first winter after dispersal while dormancy of the shoot bud is broken during the second winter.
- Physiological dormancy means that the embryo can, due to physiological causes, not generate enough power to break through the seed coat, endosperm or other covering structures. Dormancy is typically broken at cool wet, warm wet or warm dry conditions. Abscisic acid is usually the growth inhibitor in seeds and its production can be affected by light.
  - Drying; some plants including a number of grasses and those from seasonally arid regions need a period of drying before they will germinate, the seeds are released but need to have a lower moisture content before germination can begin. If the seeds remain moist after dispersal, germination can be delayed for many months or even years. Many herbaceous plants from temperate climate zones have physiological dormancy that disappears with

drying of the seeds. Other species will germinate after dispersal only under very narrow temperature ranges, but as the seeds dry they are able to germinate over a wider temperature range.

- Combinational dormancy in seeds with combinational dormancy the seed or fruit coat is impermeable to water and the embryo has physiological dormancy. Depending on the species physical dormancy can be broken before or after physiological dormancy is broken.
- Secondary dormancy is caused by conditions after the seed has been dispersed and occurs in some seeds when non–dormant seed is exposed to conditions that are not favourable to germination, very often high temperatures. The mechanisms of secondary dormancy are not yet fully understood but might involve the loss of sensitivity in receptors in the plasma membrane.

The following types of seed dormancy do not involve seed dormancy strictly spoken as lack of germination is prevented by the environment not by characteristics of the seed itself:

- Photodormancy or light sensitivity affects germination of some seeds. These photoblastic seeds need a period of darkness or light to germinate. In species with thin seed coats, light may be able to penetrate into the dormant embryo. The presence of light or the absence of light may trigger the germination process, inhibiting germination in some seeds buried too deeply or in others not buried in the soil.
- Thermodormancy is seed sensitivity to heat or cold. Some seeds including cocklebur and amaranth germinate only at high temperatures (30 °C or 86 °F) many plants that have seed that germinate in early to mid summer have thermodormancy and germinate only when the soil temperature is warm. Other seeds need cool soils to germinate, while others like celery are inhibited when soil temperatures are too warm. Often thermodormancy requirements disappear as the seed ages or dries.

Not all seeds undergo a period of dormancy. Seeds of some mangroves are viviparous, they begin to germinate while still attached to the parent. The large, heavy root allows the seed to penetrate into the ground when it falls. Many garden plants have seeds that will germinate readily as soon as they have water and are warm enough, though their wild ancestors may have had dormancy, these cultivated plants lack seed dormancy.

After many generations of selective pressure by plant breeders and gardeners dormancy has been selected out. For annuals, seeds are a way for the species to survive dry or cold seasons. Ephemeral plants are usually annuals that can go from seed to seed in as few as six weeks.

Seed Germination

Seed germination is a process by which a seed embryo develops into a seedling. It involves the reactivation of the metabolic pathways that lead to growth and the emergence of the radicle or seed root and plumule or shoot. The emergence of the seedling above the soil surface is the next phase of the plants growth and is called seedling establi-shment.

Three fundamental conditions must exist before germination can occur:

1. The embryo must be alive, called seed viability.
2. Any dormancy requirements that prevent germination must be overcome.
3. The proper environmental conditions must exist for germination.

Seed viability is the ability of the embryo to germinate and is affected by a number of different conditions. Some plants do not produce seeds that have functional complete embryos or the seed may have no embryo at all, often called empty seeds. Predators and pathogens can damage or kill the seed while it is still in the fruit or after it is dispersed. Environmental conditions like flooding or heat can kill the seed before or during germination. The age of the seed affects its health and germination ability: since the seed has a living embryo, over time cells die and cannot be replaced. Some seeds can live for a long time before germination, while others can only survive for a short period after dispersal before they die.

Seed vigour is a measure of the quality of seed, and involves the viability of the seed, the germination percentage, germination rate and the strength of the seedlings produced. The germination percentage is simply the proportion of seeds that germinate from all seeds subject to the right conditions for growth. The germination rate is the length of time it takes for the seeds to germinate. Germination percentages and rates are affected by seed viability, dormancy and environmental effects that impact on the seed and seedling. In agriculture and horticulture quality seeds have high viability, measured by germination percentage plus the rate of germination.

This is given as a per cent of germination over a certain amount of time, 90 per cent germination in 20 days, for example. 'Dormancy' is covered above; many plants produce seeds with varying degrees of dormancy, and different seeds from the same fruit can have different degrees of dormancy. It's possible to have seeds with no dormancy if they are dispersed right away and do not dry (if the seeds dry they go into physiological dormancy). There is great variation amongst plants and a dormant seed is still a viable seed even though the germination rate might be very low. Environmental conditions effecting seed germination include; water, oxygen, temperature and light. Three distinct phases of seed germination occur: water imbibition; lag phase; and radicle emergence. In order for the seed coat to split, the embryo must imbibe (soak up water), which causes it to swell, splitting the seed coat. However, the nature of the seed coat determines how rapidly water can penetrate and subsequently initiate germination.

The rate of imbibition is dependent on the permeability of the seed coat, amount of water in the environment and the area of contact the seed has to the source of water. For some seeds, imbibing too much water too quickly can kill the seed. For some seeds, once water is imbibed the germination process cannot be stopped, and drying then becomes fatal. Other seeds can imbibe and lose water a few times without causing ill effects, but drying can cause secondary dormancy.

Inducing Germination

A number of different strategies are used by gardeners and horticulturists to break seed dormancy. Scarification which allows water and gases to penetrate into the seed, include methods that physically break the hard seed coats or soften them by chemicals.

Means of scarification include soaking in hot water or poking holes in the seed with a pin or rubbing them on sandpaper or cracking with a press or hammer. Soaking the seeds in solvents or acids is also effective for many seeds. Sometimes fruits are harvested while the seeds are still immature and the seed coat is not fully developed and sown right away before the seed coat become impermeable. Under natural conditions seed coats are worn down by rodents chewing on the seed, the seeds rubbing against rocks (seeds are moved by the wind or water currents), by undergoing freezing and thawing of surface water, or passing through an animal's digestive tract.

In the latter case, the seed coat protects the seed from digestion, while often weakening the seed coat such that the embryo is ready to sprout when it gets deposited (along with a bit of fertilizer) far from the parent plant. Microorganisms are often effective in breaking down hard seed coats and are sometimes used by people as a treatment, the seeds are stored in a moist warm sandy medium for several months under non–sterile conditions. Stratification also called moist–chilling is a method to break down physiological dormancy and involves the addition of moisture to the seeds so they imbibe water and then the seeds are subject to a period of moist chilling to after–ripen the embryo. Sowing outside in late summer and fall and allowing to overwinter outside under cool conditions is an effective way to stratify seeds, some seeds respond more favourably to periods of oscillating temperatures which are part of the natural environment.

Leaching or the soaking in water removes chemical inhibitors in some seeds that prevent germination. Rain and melting snow naturally accomplish this task. For seeds planted in gardens, running water is best—if soaked in a container, 12 to 24 hours of soaking is sufficient. Soaking longer, especially in stagnant water that is not changed, can result in oxygen starvation and seed death. Seeds with hard seed coats can be soaked in hot water to break open the impermeable cell layers that prevent water intake.

Other methods used to assist in the germination of seeds that have dormancy include prechilling, predrying, daily alternation of temperature, light exposure, potassium nitrate, the use of plant growth regulators like gibberellins, cytokinins, ethylene, thiourea, sodium hypochlorite plus others. Some seeds germinate best after a fire, for some seeds fire cracks hard seed coats while in other seeds chemical dormancy is broken in reaction to the presence of smoke, liquid smoke is often used by gardeners to assist in the germination of these species.

Origin and Evolution

The origin of seed plants is a problem that still remains unsolved. However, more and more data tends to place this origin in the middle Devonian. The description in 2004 of the proto–seed Runcaria heinzelinii in the Givetian of Belgium is an indication of that ancient origin of seed–plants. As with modern ferns, most land plants before this time reproduced by sending spores into the air, that would land and become whole new plants. The first "true" seeds are described from the upper Devonian, which is probably the theater of their true first evolutionary radiation. The seed plants progressively became one of the major elements of nearly all ecosystems.

GERMINATION

When the pod of the bean is developing, the embryo in the seed is being fed by the parent and visibly grows until ripeness is attained. The young plant then assumes a dormant or resting state within the seed without showing any signs of life.

Under certain conditions, however, the plantlet begins to wake up, and soon escapes from its protective coat to lead a separate and independent life. This awakening from a resting condition to a state of active growth is called germination and is dependent upon an adequate supply of

- Water
- Heat
- Air or oxygen.

It is also essential, of course, that the plantlet in the seed must be alive. The exact nature of the dormant state of seeds is not understood, but in old seeds and those which are gathered in an immature condition or badly stored the embryo is often weakened or actually dead; in the latter case no germination is possible.

The exact length of time which seeds may be kept before death of the embryo takes place has never been satisfactorily determined; it varies with the species of the seed, its ripeness and composition, and also with the method of storage. In the case of most farm and garden seeds kept in the ordinary way, few of them are found capable of growth after ten years, and a large number die in two or three years. For present purposes it will suffice merely to mention that age is a determining factor in the germination of seeds. Water is necessary is well known, as beans may be kept indefinitely in a sack or drawer at various temperatures and with access to air without germination taking place.

When placed in moist ground, or between damp blotting-paper, they absorb water very readily. This is most easily observed when beans are soaked for twelve hours in a dish containing water. The water is transmitted through all parts of the coat, but much more quickly and easily through the micropyle and the line of softer material which runs the whole length of the centre of the hilum.

It is rapidly brought into contact with the part of the embryo which grows first, namely, the radicle. The soft spongy thicker part of the inside of the testa lying beneath the hilum stores up a considerable amount of water for the benefit of the developing plant, and the whole of the embryo and the seed-coat absorb water and become softer and larger in consequence; it is only after this swelling has happened that a bean begins to show any signs of germination.

The need of an adequate temperature for germination is a matter of common knowledge among those accustomed to sow seeds. If soaked beans are placed in the ground in midwinter they show little or no signs of waking from their dormant condition, yet when placed under a glass on damp blotting-paper indoors, the radicle makes its exit from the seed in a few days.

Seeds differ in the temperature which is necessary to induce them to germinate, the embryos in some commence to extend their radicles and push their way through the seed-coat even if just kept above freezing point; others require a temperature of 9° or 10°C to start growth.

If attempts are made to grow beans at 45°C it will be found to be too hot, and they make little or no progress. Between this high temperature at which growth appears impossible, and the freezing point where the development of the embryo of the bean is also suspended, there is a temperature at which the embryo makes the most rapid progress, and emerges from the seed-coat in the shortest possible time; this most favourable temperature is about 28°C, both above it and below it the germination of the bean is retarded. The supply of fresh air is also an essential condition for growth of the young plant from the bean seed, but the evidence for its need is not so manifest or so generally recognised as the necessity for moisture and warmth.

It will be found, however, when beans are placed in a flask or bottle containing carbon dioxide or hydrogen gas they refuse to germinate, even when they are supplied with a proper amount of water.

The peculiar extension or growth of the parts of the interior of the bean seed, and the fact that a suitable supply of water, air and heat is necessary for the manifestation of these changes, suggests to us that we are dealing with a living structure.

This becomes all the more«apparent when we observe that the oxygen of the air is absorbed, and in its place carbon dioxide is given off into the surrounding air, for this is what happens in the breathing of a living animal. Carbon dioxide is produced when beans germinate.

Place twenty soaked beans in a wide-mouthed bottle, and cork them up after showing that a match burns freely in the bottle. Leave them in a warm place for twenty-four hours, and try if a match will now burn in the bottle. The carbon dioxide gas can be poured out into a beaker containing lime water; on shaking, its presence is proved by the lime water becoming 1 milky ' owing to the precipitation of carbonate of lime. The particular use of the water, heat and air to the plant we cannot at present discuss.

Without water the embryo would have little chance of becoming free from the tough and hard seed-coat surrounding it; water softens the latter, and makes it more easily torn by the extending radicle and plumule. In the early stages of the life of the bean plant, from the commencement of germination up to the time when the first green leaves are unfolded, the development and building up of the elongating rootlet and shoot depend upon the thick cotyledons.

At first the latter are thick and fleshy, but as the radicle and plumule grow the cotyledons become softer and thinner, ultimately shrivelling considerably. The cotyledons are leaves, the interior of which is packed with food for the rest of the growing embryo, and a large amount of the water absorbed by the seed is used for the purpose of dissolving the nutrient material in them, and carrying it from them to the various parts of the root and shoot of the young plant where growth is going on the cotyledons are esential to the development of the root and shoot of the embryo by cutting them off as soon as the two latter parts have emerged from the seed-coat. Try separating one cotyledon and then two at various stages of development, and see if the axis (root and shoot) can be made to develop without them.

The growth should be allowed to continue some time in order to obtain well-marked effects. Not only do the changes observed innhe embryo of a germinating bean point to the conclusion that it is a living structure and like an animal dependent on a proper supply of water, heat and air for the manifestation of its life, but the parts of a young bean plant after emerging from the seed soon give evidence of the possession of peculiarities which are associated with life. When put in the ground, the radicle, in coming out of the seed, turns straight downwards and continues to grow in this direction. This is the case no matter in what position the seed is placed. If, after germination has commenced, it is taken and replanted with the primary root pointing to the surface of the soil, the tip of the root soon begins to curve downwards again, and will maintain this course until again disturbed.

The plumule behaves in exactly the opposite manner; after emerging from the seed-coat its bent tip grows upwards and away from the root; if the seed is reversed and replanted the plumule begins to curve in such a manner that its tip is driven upwards towards the surface of the ground.

That these peculiarities are somehow connected with life is clear, as dead embryos show no such behaviour. Sow soaked beans in a flower pot or box filled with ordinary garden soil placing them in various positions in it, some laid on the flat side, some with the hilum directed upwards, and others with the hilum downwards. Allow them to grow in a warm place: take them up as soon as signs of germination are noticed, and observe the direction the root and shoot have taken. The peculiar tendency for the root always to go downwards and stem upwards can be investigated by sowing beans in ordinary garden soil and afterwards reversing them.

To avoid error all should be taken up, and then placed again in the soil in various positions—some as they were, a few with their roots and stems reversed, and others laid in a horizontal position. They may be re-examined at the end of a week. When the roots have extended about half an inch take two seeds and suspend them by means of thread side by side in a bottle with their roots downwards and stem upwards. The bottle should contain a little water to keep the air damp.

When the roots have grown about two inches reverse one of the seeds so that its root points upwards tad stem downwards. Notice that the tip of the root of the reversed seed in about twelve hours begins to turn downwards, while the plumule more slowly bends in such a way as to assume the position it had before it was reversed.

The bottle should be placed in a dark box or cupboard to avoid the influence of light on the plant, and fresh air should be blown into the bottle twice a day. Although seeds vary almost indefinitely in regard to size and shape they are similar to the bean in so far as they all contain a young plant packed away within the seed-coats.

In this essential feature all seeds agree with few exceptions, and it is on account of the existence of a young plant within them that they are of use in the raising of crops or plants. The manner, however, in which the embryo is arranged, and the relative size and appearance of its various parts, differ considerably in seeds; moreover, the growth during and after germination is not the same in all cases. A few of the more important and common variations in these respects must be noticed.

THE COMMON BEAN

A broad bean is one of the largest seeds met with in ordinary farm or garden practice, and as its parts are all sufficiently large to be observed without the special aid of anything more than an ordinary pocket lens, it is especially fitted for study. When a nearly ripe pod of a broad bean plant is opened, each seed within it is found attached to. The inside by means and it is through this stalk that all the nourishment passes from the parent to enable the young seed to develop. A1 first the pod exists in a rudimentary form in the centre of a flower and its parts and contents are very small; they are nevertheless readily seen with a pocket lens. After the fading of the flower, the pod and seeds within it grow larger and larger at the expense of food supplied by the rest of the plant, and ultimately when ripe the funicles wither and dry up, and the seeds become detached from the parent which has produced them. When dry and ripe each bean seed is hard, with an uneven surface, but its internal construction cannot be clearly examined in this condition.

On soaking in water for twelve hours, however, it becomes softer, and the parts can then be easily investigated. The outside, which is a pale buff colour, is smooth, and has at one end a narrow elongated black scar called the hilum of the seed. It is known popularly as the ' eye' of the bean, and marks the place where the broad end of the funicle separated from the seed when it ripened in the pod. Quite close to one end of the hilum is a very minute hole known as the micropyh, easily seen with a lens, and through which water oozes out usually accompanied by bubbles of air when soaked beans are squeezed between the finger and thumb.

This opening communicates with the interior of the seed, and is the only one it possesses. The rest of the seed after the testa is removed, is of oval flattened shape similar to the complete bean, and is divisible into two large fleshy halves called cotyledons which, however, are not completely separate from each other, but connected at the side with a conical projecting body, one end of which is found to fit into a hollow cavity in the seed-coat exactly opposite the micropyle; the other end is bent and turned inwards between the fleshy cotyledons.

The extent and shape of this small curved structure is most easily observed when one of the cotyledons is removed completely; it remains attached to the other. Soak some broad beans in water and keep them in a warm place all night. Examine them next day and make drawings of the various parts seen both before and after stripping off the testa. Observe the relative position of the parts of the embryo in reference to each other and to the seed-coat.

Examine and compare the structure of the following seeds after soaking in the same way:—Pea, scarlet runner beans, vetches, and red clover. The bean seed contains nothing more than what has already been described; the nature and relationship of its component parts only become intelligible when the seed is placed in the ground or maintained under certain conditions, and allowed to grow.

When growth commences the lower end of the small curved structure elongates and breaks its way through the coat of the seed at a point very close to the micropyle, but not, as often erroneously stated, through the micropyle itself. It soon assumes the form and is recognised as a root of a young bean plant The upper bent half, which lies between the cotyledons, also pushes its way out of the same opening in the seed-coat and develops into a stem, from the tip of which leaves are gradually unfolded.

It is thus seen that the seed of a broad bean is a packet containing a bean plant in a rudimentary condition. This plantlet is callai an embryo, and the portion of it which becomes root and stem is its primary axis.

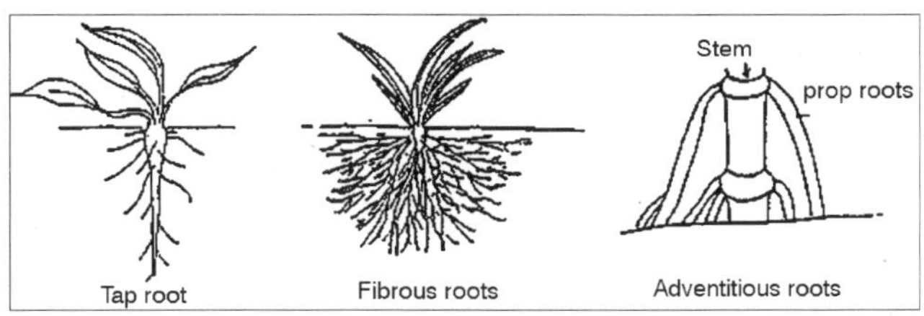

Fig. Root System

The part of the primary axis which is below the point where the cotyledons are attached consists of a very small piece of stem, the hypocotyl at the end of which is the radicle or primary root. Where the stem ends and the root begins cannot be determined in the bean seedling without the aid of the microscope and examination of the internal structure of the axis of the plant. The curved end of the primary axis above the cotyledons is the plumule of the embryo, and consists of a very short piece of stem, the tpicotyl on the top of which is a bud.

From the latter is derived the ordinary stem which comes above ground with its green leaves and flowers. In the early stages of the growth of the embryo from the seed the hypocotyl grows very little. The upper part of the stem bearing the plumule comes out of the seed bent and it maintain this curved shape for some time after emerging. By this behaviour the delicate leaves of the plumule are protected from injury during their progress upwards when a seed is sown in earth or sand. Fold up some soaked beans in two thicknesses of white flannel made damp, and place them on a plate. Cover them with another plate placed upside down, and leave them in a warm room.

Examine them twice a day, leaving them, exposed to the fresh air for a few minutes each time, and keeping the flannel damp, not wet. When they sprout notice the place where the radicle has come out of the seed-coat. Let some grow till the radicle and plumule are well out of the seed, and compare the various parts of the sprouted seeds with unsprouted ones.

SEED OF WHITE MUSTARD

The seed of white mustard (Brassica alba Boiss.) contains an embryo which like that of the bean consists of a radicle, plumule and two cotyledons; the latter, which are folded together, are relatively thinner than those of the bean and deeply notched. The radicle is bent round and lies in the fold of the cotyledons, between which is the very small, almost invisible plumule. On germination the cotyledons, instead of remaining within the seed-coat and below the ground as in the case of the broad bean, escape from the enclosing coats altogether and grow up out of the ground, enlarging at the same time, and becoming green like ordinary leaves.

They are the first ' smooth' leaves of the seedling mustard plant. After a short time the plumule grows up from between the cotyledons and forms a stem upon which are gradually unfolded the ordinary divided ' rough' leaves. Soak white mustard seeds, and examine their structure, noting especially the way in which the embryos are packet in them.

Allow some to germinate and grow for a week or more on damp flannel, and examine them in various stages of development, noting the notched cotyledons with small. The term hypogean is applied to cotyledons which remain below ground, those coming above being epigean, the relative amount of growth in the hypocotyl and epicotyl determining their position. If the hypocotyl grows vigorously during or after germination the cotyledons are forced above ground; when only the epicotyl grows the plumule is lifted up above the soil, but the cotyledons remain below where the seed is placed. In the broad bean the hypocotyl is very short, and the point where it ends and the root begins is not clearly denned. In a mustard seedling, however, the point of separation between the root and stem is somewhat swollen and readily distinguished. All plants whose embryos, like those of the bean and mustard, possess two cotyledons, are known as Dicotyledons they form a very large, well-marked class of the flowering or seed-bearing plants. The seeds contain within their coats nothing but an embryo plant, which depends for the development of its root and shoot upon the substances stored up in some part of its body, its cotyledons chiefly.

This is true even in the case of seeds like those of white mustard, in which the cotyledons of the embryo are comparatively thin. There are, however, a number of plants, such as the ash, mangel and potato, which, although belonging to the Dicotyledons, have seeds in which there are stores of food inside the seed-coat, but free from the embryo and its cotyledons. Such separate reserve-food is stored in that part of the seed known as the endosperm or 'albumen,1 and seeds in which it is present are called endospermous or albuminous seeds. Those like the bean, pea, and vetch, mustard, and turnip, which have no separate reserve-food, are known as exendospermous or exalbuminous seeds.

Take out a seed from the fruit of the ash tree in autumn; carefully cut thin shavings from the flat side of the seed, stafting about the middle of the seed and cutting towards the narrow end. Note the white embryo with its well-marked radicle, hypocotyl and two flat cotyledons lying within the semi-transparent endosperm. Some of the most commonly occurring endospermous seeds will be found to have embryos within them which are not dicotyledonous, and whose structure is in many respects. A good example is met with in the onion.

WHEAT GRAIN

A wheat grain, which may be taken as an example, is not a seed, but a kind of nut with a single seed within it. The seed grows in such a way as to completely fill up the interior of the nut, and become practically united with its inside wall. The embryo occupies only a small part of the grain, the rest being taken up by the floury endosperm of the seed. The embryo is easily seen at the base of a soaked grain on the side opposite the furrow.

When removed it has the appearance. The part of it which lies close up to the endosperm is a flattened somewhat fleshy shield-shaped structure called the scutcllum attached to the front of the scutellum is the plumule, consisting of a bud formed of an extremely short stem, upon which are sheath-like leaves enclosing each other.

The embryo generally possesses five roots, one primary and two pairs of secondary. They are all completely enclosed by a sheath which continuous wjtn the scutellum, and are consequently not visible from outside; their position, however, is marked by projecting bosses. The sheath round the roots is termed the coleorhiza, and when germination takes place it expands and bursts the coats of the grain, the roots about the same time breaking through the enclosing coleo-rhiza.

When a wheat grain is sown in the ground it remains there, but the plumule grows upwards, its first leaf, the coleoptile, appearing above the soil as a single pale tube-like structure; from a slit in the tip of the latter the first flat green blade soon appears, and is followed by a succession of single green leaves, the younger ones growing from within the older ones in regular order. With a sharp knife or razor cut through from back to front, so as to divide the grain into two longitudinal halves, and note the floury endosperm and the shape and parts of the divided embryo. Place a folded sheet of damp blotting paper on a plate, sow some soaked wheat grains on it, and cover with a tumbler. The grains will germinate. Watch their development up to the time the first green leaf appears, taking out the embryo and examining it at different stages of its growth. There is difference of opinion as to which part of the embryo is to be considered the cotyledon.

Soine authorities regard the scutellum as the cotyledon, while others give this name to the coleoptile or first sheathing leaf which comes above ground, and which has no green blade. Others, again, consider that the first sheathing leaf is an extension of the scutellum, and the two combined is therefore the cotyledon.

In any case, there is only one cotyledon present, and wheat therefore belongs to the class of monocotyledonous plants. During the growth of the embryo of a wheat grain, it will be noticed that the endosperm becomes soft and decreases in quantity as the roots and plumule expand and develop; the endosperm is the food upon which the young plant depends during the early staggs of its life, the scutellum acting as a structure for changing, absorbing, and transferring this reserve-food to the growing parts which need it. Remove the embryos from well-soaked grains, and grow them without the endosperm on damp blotting-paper.

Allow ordinary uninjured grains to grow with them. Both the embryos in the grains and those removed from the grains develop, but there is a great difference in the results after a few days. The store of reserve-food on which the young plant depends for its early development is sufficient to enable it to form a root, stem, and several leaves, as is evident when seeds are allowed to germinate upon damp flannel or blotting-paper, from which nothing but water is absorbed. No food-materials or manures are needed for this primary development, and seeds germinate and the seedlings grow for a considerable time as well in poor soil or sand as in good rich ground.

As soon as the reserve store is exhausted hunger becomes apparent, and unless the plants are then supplied with suitable nutriment from the soil and air, and are also placed under conditions favourable for growth, weakness and death are likely to occur. Among the larger seeds, such as beans and peas, where there is an abundant store of reserve-food, the young seedlings begin to manufacture food for themselves from materials absorbed from the soil and air, long before their reserve is exhausted.

In small seeds the reserve is sometimes almost consumed before the roots and leaves are sufficiently developed to carry on their work properly, in which cases a more or less temporary starvation and check to growth ensues. Especially does this happen when seeds are sown too deeply, for a large amount of food is then used in the production of a stem long enough to lift the leaves up into the air.

ONION

The seed is black, somewhat oval in outline, with one side convex, the other almost flat. Each contains within it endosperm and an embryo which lies curled up inside in the form seen. When germination commences, the curved part imbedded in the middle of the endosperm grows and forces the end of the embryo out of the seed. From this exposed end, which is the radicle, a straight, slender, primary root develops. The part of the young seedling which extends from the root into the interior of the seed, grows very rapidly at first, at the same time assuming a sharply bent outline.

It comes above ground in the form of a close loop but on further growth the end within the seed is pulled out of the soil, and grows up in the air. The tip within the seed changes and absorbs the endosperm, and usually remains there until all the nutrient material has been transferred from it to the various centres of growth in the young plant. After the food-reserve is exhausted the tip withers and becomes free from the seed coat.

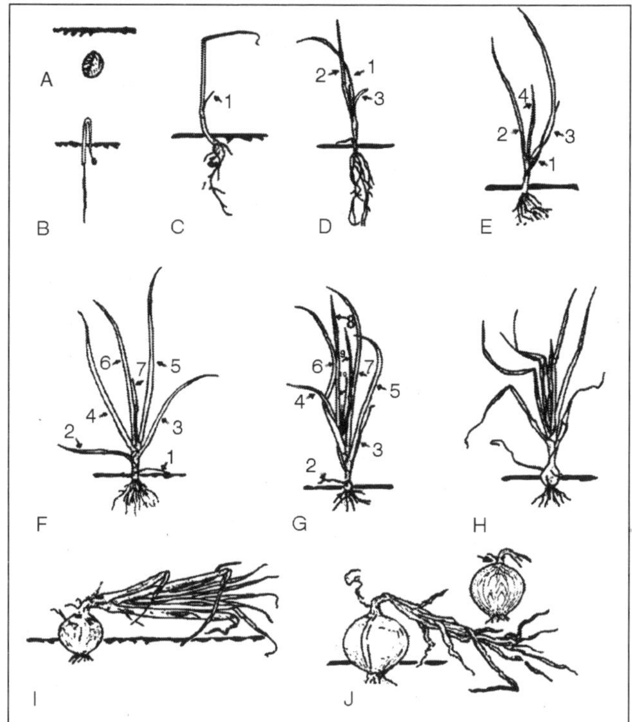

Fig. Germination of Onion

In loose soils the latter is pulled above ground before the endosperm is exhausted, and remains on the end of the tip for some time. In other cases where the soil is damper and of a stiffer nature the seed-coat remains below ground altogether. The curved part of the embryo which comes above ground is a leaf. It is the cotyledon of the embryo, and is in reality a thin hollow leaf like those of the full grown onion plant: within it is the plumule, which consists of a series of hollow conical leaves arranged one inside the other.

After one leaf emerges others soon follow, the younger ones coming out in regular order through slits in the sides of those immediately older than themselves. Soak fresh onion seeds in water for a few hours. With a razor cut through some parallel to their flat sides in order to show the embryo within.

Sow others in damp blotting-paper; allow them to germinate and the seedlings to develop; make observations of them at different stages of growth. Watch the germination of seeds sown in boxes or pots containing ordinary garden soil. Plants whose embryos possess only one cotyledon are known as Monocotyledons, and form the second large class of seed-bearing plants. Few of the representatives of this class with which we are ordinarily familiar have true seeds large enough for examination. The onion is probably one of the best commonly occurring examples which may be considered typical of the monocotyledons and easily obtainable. To this class, however, belong all the grasses, but their seeds and embryos are so different in many respects from those of the onion that it is necessary to examine one of them in detail.

PRIMARY AND SECONDARY BOOTS

It was noticed, when dealing with the bean seedling, that its two ends always grow in opposite directions; the plantlet can be considered as an elongated axis, one end of which bears the leaves and invariably comes above ground, while the opposite end never bears leaves, and persistently follows the plumb-line downwards. The descending part is known as the root. As will be pointed out later, all roots do not behave in this manner, and it is to be specially noted that many of the underground parts of plants are not roots; the exceptions, however, may be left for future consideration. The first or primary root which the bean plant possesses is merely an extension of the radicle of the embryo which exists nrithin the seed.

Soon after making its exit from the seed, it takes a downward course, and elongates by growth taking place near its tip. Another point to be noted is that the lateral roots do not arise as up-growths on the surface of the primary root, but come from within it, and are described as endogenous. The slits which they make in the substance of the primary root, and through which they emerge, can be readily seen in a bean seedling.

The secondary lateral roots are connected with its central more solid core; the three lowest ones, although they have just begun to grow, have not yet penetrated the outer layer of the root, and would not be visible on the outside of the latter. This mode of origin is characteristic of lateral roots generally wherever they are met with. Many dicotyledonous plants have roots similar to those of the bean plant.

When, as in this case, the primary one continues to grow, keeping distinctly larger than the lateral ones, it is called a tap root. Very good examples are met with among cultivated plants in the carrot, mangel, red clover, and mustard in shep-herd's-purse, poppy, and many other weeds, as well as in most broad-leaved trees. A number of plants have swollen fleshy roots in which food materials are stored for future use; they are described as tuberous roots, and must be distinguished from tubers, which are fleshy underground stems. To designate the different forms of thickened roots various special terms are in use.

The typical carrot root is conical; that of the turnip napiform. The root of the radish is spoken of as fusiform. In some instances the primary root is soon rivalled in size by its branches; it may even cease growth altogether. Such plants, on being pulled out of the ground, exhibit a bunch of slender roots, the chief of which are all much the same diameter and length; roots of this character are described as fibrous and are well exemplified in common groundsel and grasses.

ADVENTITIOUS ROOTS OF MONOCOTYLEDONOUS PLANTS

The roots of monocotyledonous plants differ in their development from those of dicotyledons. The single primary root of the onion, lasts but a short time, and is succeeded by others which do not arise as branches upon the primary one, but spring from the very short stem of the plant. Roots which arise on stems and leaves, or on various parts of the roots of plants, but not in acropetal succession are described as adventitious. They are very common upon all monoco-tyledonous plants of the farm and garden, and may be considered the chief ones which such plants possess.

In wheat, the embryo within the grain possesses three roots; in barley, five or six. Young barley plant show- use during the earlier stages of growth. Although common upon monocotyledons adventitious roots are not confined to this class. Examples are abundant upon many kinds of dicotyledonous plants.

Good instances are met with on the underground stems of mint, potato and hop, and on the runners of the strawberry, stems of creeping crowfoot and white clover, as well as many others. They are generally produced at the joints where leaves grow upon the stem, and may arise in some plants from internal causes apart from any external influences; in other instances their development depends upon contact of the stem with water or moist soil.

Almost all parts of certain plants may be made to produce them, and the propagation of plants Buch as gooseberries, currants, roses and hops by means of slips and cuttings depends upon their development. Pieces of stem cut off just below a leaf, and placed in moist earth soon develop adventitious roots near the cut end. Advantage is taken of their formation in the propagation of plants by means of layers.

Examine the roots upon any cuttings or slips which can be obtained, and observe whether they arise on the cut surface or at a point some distance away from it. Usually adventitious roots are thin and fibrous, but those of the dahlia are tuberous. The complete root-systems of plants vary enormously in extent, but in all cases the total length is much greater than is usually anticipated.

That of an ordinary oat plant, although not spreading through more than a cubic yard or two of soil, measured in one instance over one hundred and fifty yards in length. A tree uprooted by the wind exposes to view a small number of stout roots similar to the thicker branches of the crown, and from these are given off a larger number of a finer texture. By far the greater bulk, however, which the tree possessed remain in the ground in the form of extremely fine rootlets or fibrils extending outwards generally as far, or a little farther, than the branches and leaves of the tree, but in some instances to much greater distances.

Not only do the roots grow out horizontally and near the surface of the soil, but they extend downwards as well. In isolated instances, where an adequate supply of air has been maintained along open cracks and fissures, roots have been found to descend many yards into the ground, but in general the roots of the tallest trees rarely go down more than seven or eight feet. The want of air and presence of noxious substances in the lower regions of the soil checks further growth in this direction.

With many plants almost every cubic inch of soil immediately beneath their shade contains fine delicate rootlets, and the extent of their root-branching is very rarely realised on account of the ease with which these hair-like fibrils are broken off when the plant is pulled up or disturbed. Many forest trees have a natural habit of sending their roots several feet into the soil Examples of fruit-trees belonging to this class, and requiring a deep soil for proper growth, are the cherry and wild pear. Some trees, such as larch, keep their root-system nearer the surface, spreading out more horizontally in the ground.

The quince, used as a stock on which to graft pears, has roots which remain in the upper regions of the soil. Similar surface-rooting habit is very marked in the ' Paradise' apple on which apples are grafted, and in Prunus Mahaleb upon which cherries are often grafted. The root-system of wheat penetrates more deeply than that of barley; the mangel sends its fine rootlets more extensively into the deeper layers of the soil than cabbage or turnip red clover roots more deeply than white clover, and almost all plants have distinct and peculiar habits in this respect.

The character and extent of root-development is not, however, altogether dependent upon the species of the plant concerned, but is materially influenced by external circumstances, such as the texture of the soil, and the amount of water in it. In deep open soils and loose sands the root-system of a plant is much larger than that of a similar plant grown in compact heavy ground. In soils which are not water-logged, increase of moisture up to a certain extent increases the branching of the root, and excellent examples of the influence of water, coupled with good air supply, are seen in well managed plants in pots, and also among plants growing near wells and in drain pipes; the latter in some instances become completely blocked by the large number of fine rootlets of trees growing in their neighbourhood. The root-system is also considerably modified by the totyl amount and kind of the manures or food-materials present in the soil. Up to a certain point an increase of nutrient substances increases root-development; an excess hinders it. Mutilation influences the development of the root-system. If the growing-point of the tap root of a cabbage or tree is cut off its further elongation is prevented, but the secondary roots make up for the loss by more vigourous growth and often many adventitious roots arise near the cut end. In order to properly cultivate plants of all kinds it is very important to study the habit or manner of branching of their roots and the relative proportions of the thick tap and secondary roots to the fine ramifications to which they give rise and which spread through the soil in all directions.

The proportion of root-system below ground to the branches and leaves above is also worthy of attention. The adaptability of plants to the various kinds of soils, their need of water, the cultivation which the ground should have and the rational application of manures to the plant are best understood and appreciated after a careful study of these points. Tap-rooted crops, such as sugar-beet, mangel, carrot, and parsnip, require the soil to be well-worked to some considerable depth. Surface-rooting plants, such as barley, can be grown upon comparatively thin soils.

The same is true of pears grafted on the quince and apples on the Paradise stock, and such plants respond more quickly to, and are more benefited by, top-dressings of soluble manures than plants with a deeply-penetrating root-system.

FACTORS AFFECTING SEED GERMINATION

Seed germination depends on both internal and external conditions. The most important external factors include temperature, water, oxygen and sometimes light or darkness. Various plants require different variables for successful seed germination, often this depends on the individual seed variety and is closely linked to the ecological conditions of a plant's natural habitat.

For some seeds, their future germination response is affected by environmental conditions during seed formation; most often these responses are types of seed dormancy.

WATER

Water is required for germination. Mature seeds are often extremely dry and need to take in significant amounts of water, relative to the dry weight of the seed, before cellular metabolism and growth can resume.

Most seeds need enough water to moisten the seeds but not enough to soak them. The uptake of water by seeds is called imbibition, which leads to the swelling and the breaking of the seed coat. When seeds are formed, most plants store a food reserve with the seed, such as starch, proteins, or oils. This food reserve provides nourishment to the growing embryo. When the seed imbibes water, hydrolytic enzymes are activated which break down these stored food resources into metabolically useful chemicals.

After the seedling emerges from the seed coat and starts growing roots and leaves, the seedling's food reserves are typically exhausted; at this point photosynthesis provides the energy needed for continued growth and the seedling now requires a continuous supply of water, nutrients, and light.

OXYGEN

Oxygen is required by the germinating seed for metabolism. Oxygen is used in aerobic respiration, the main source of the seedling's energy until it grows leaves. Oxygen is an atmospheric gas that is found in soil pore spaces; if a seed is buried too deeply within the soil or the soil is waterlogged, the seed can be oxygen starved. Some seeds have impermeable seed coats that prevent oxygen from entering the seed, causing a type of physical dormancy which is broken when the seed coat is worn away enough to allow gas exchange and water uptake from the environment.

TEMPERATURE

Temperature affects cellular metabolic and growth rates. Seeds from different species and even seeds from the same plant germinate over a wide range of temperatures. Seeds often have a temperature range within which they will germinate, and they will not do so above or below this range.

Many seeds germinate at temperatures slightly above room–temperature 60–75 °F (16–24 °C), while others germinate just above freezing and others germinate only in response to alternations in temperature between warm and cool. Some seeds germinate when the soil is cool 28–40 °F (–2—4 °C), and some when the soil is warm 76–90 °F (24–32 °C). Some seeds require exposure to cold temperatures (vernalization) to break dormancy. Seeds in a dormant state will not germinate even if conditions are favourable.

Seeds that are dependent on temperature to end dormancy have a type of physiological dormancy. For example, seeds requiring the cold of winter are inhibited from germinating until they take in water in the fall and experience cooler temperatures.

Four degrees Celsius is cool enough to end dormancy for most cool dormant seeds, but some groups, especially within the family Ranunculaceae and others, need conditions cooler than –5 C.

Some seeds will only germinate after hot temperatures during a forest fire which cracks their seed coats; this is a type of physical dormancy.

Most common annual vegetables have optimal germination temperatures between 75-90 °F (24–32 °C), though many species (*e.g.* radishes or spinach) can germinate at significantly lower temperatures, as low as 40 °F (4 °C), thus allowing them to be grown from seed in cooler climates. Suboptimal temperatures lead to lower success rates and longer germination periods.

LIGHT OR DARKNESS

Light or darkness can be an environmental trigger for germination and is a type of physiological dormancy. Most seeds are not affected by light or darkness, but many seeds, including species found in forest settings, will not germinate until an opening in the canopy allows sufficient light for growth of the seedling.

Scarification mimics natural processes that weaken the seed coat before germination. In nature, some seeds require particular conditions to germinate, such as the heat of a fire (*e.g.*, many Australian native plants), or soaking in a body of water for a long period of time. Others need to be passed through an animal's digestive tract to weaken the seed coat enough to allow the seedling to emerge.

DORMANCY

Some live seeds are dormant and need more time, and/or need to be subjected to specific environmental conditions before they will germinate. Seed dormancy can originate in different parts of the seed, for example, within the embryo; in other cases the seed coat is involved.

Dormancy breaking often involves changes in membranes, initiated by dormancy–breaking signals.

This generally occurs only within hydrated seeds. Factors affecting seed dormancy include the presence of certain plant hormones, notably abscisic acid, which inhibits germination, and gibberellin, which ends seed dormancy. In brewing, barley seeds are treated with gibberellin to ensure uniform seed germination for the production of barley malt.

SEEDLING ESTABLISHMENT

In some definitions, the appearance of the radicle marks the end of germination and the beginning of "establishment", a period that ends when the seedling has exhausted the food reserves stored in the seed.

Germination and establishment as an independent organism are critical phases in the life of a plant when they are the most vulnerable to injury, disease, and water stress. The germination index can be used as an indicator of phytotoxicity in soils. The mortality between dispersal of seeds and completion of establishment can be so high that many species have adapted to produce huge numbers of seeds.

Germination Rate

In agriculture and gardening, the germination rate describes how many seeds of a particular plant species, variety or seedlot are likely to germinate. It is usually expressed as a percentage, *e.g.*, an 85 per cent germination rate indicates that about 85 out of 100 seeds will probably germinate under proper conditions. The germination rate is useful for calculating the seed requirements for a given area or desired number of plants.

Dicot Germination

The part of the plant that first emerges from the seed is the embryonic root, termed the radicle or primary root. It allows the seedling to become anchored in the ground and start absorbing water. After the root absorbs water, an embryonic shoot emerges from the seed. This shoot comprises three main parts: the cotyledons (seed leaves), the section of shoot below the cotyledons (hypocotyl), and the section of shoot above the cotyledons (epicotyl). The way the shoot emerges differs among plant groups.

Epigeous

In epigeous (or epigeal) germination, the hypocotyl elongates and forms a hook, pulling rather than pushing the cotyledons and apical meristem through the soil. Once it reaches the surface, it straightens and pulls the cotyledons and shoot tip of the growing seedlings into the air. Beans, tamarind, and papaya are examples of plants that germinate this way.

Hypogeous

Another way of germination is hypogeous (or hypogeal), where the epicotyl elongates and forms the hook. In this type of germination, the cotyledons stay underground where they eventually decompose. Peas, for example, germinate this way.

Monocot Germination

In monocot seeds, the embryo's radicle and cotyledon are covered by a coleorhiza and coleoptile, respectively. The coleorhiza is the first part to grow out of the seed, followed by the radicle. The coleoptile is then pushed up through the ground until it reaches the surface. There, it stops elongating and the first leaves emerge.

Precocious Germination

While not a class of germination, precocious germination refers to seed germination before the fruit has released seed. The seeds of the green apple commonly germinate in this manner.

Pollen Germination

Another germination event during the life cycle of gymnosperms and flowering plants is the germination of a pollen grain after pollination. Like seeds, pollen grains are severely dehydrated before being released to facilitate their dispersal from one plant to another. They consist of a protective coat containing several cells (up to 8 in gymnosperms, 2–3 in flowering plants). One of these cells is a tube cell.

Once the pollen grain lands on the stigma of a receptive flower (or a female cone in gymnosperms), it takes up water and germinates. Pollen germination is facilitated by hydration on the stigma, as well as by the structure and physiology of the stigma and style.

Pollen can also be induced to germinate in vitro (in a petri dish or test tube). During germination, the tube cell elongates into a pollen tube. In the flower, the pollen tube then grows towards the ovule where it discharges the sperm produced in the pollen grain for fertilization. The germinated pollen grain with its two sperm cells is the mature male microgametophyte of these plants.

SELF–INCOMPATIBILITY

Since most plants carry both male and female reproductive organs in their flowers, there is a high risk of self–pollination and thus inbreeding. Some plants use the control of pollen germination as a way to prevent this self–pollination. Germination and growth of the pollen tube involve molecular signaling between stigma and pollen. In self–incompatibility in plants, the stigma of certain plants can molecularly recognize pollen from the same plant and prevent it from germinating.

Spore Germination

Germination can also refer to the emergence of cells from resting spores and the growth of sporeling hyphae or thalli from spores in fungi, algae and some plants. Conidia are asexual reproductive spores of fungi which germinate under specific conditions. A variety of cells can be formed from the germinating conidia. The most common are germ tubes which grow and develop into hyphae. Another type of cell is a conidial anastomosis tube (CAT); these differ from germ tubes in that they are thinner, shorter, lack branches, exhibit determinate growth and home towards each other. Each cell is of a tubular shape, but the conidial anastomosis tube forms a bridge that allows fusion between conidia.

Resting Spores

In resting spores, germination that involves cracking the thick cell wall of the dormant spore. For example, in zygomycetes the thick–walled zygosporangium cracks open and the zygospore inside gives rise to the emerging sporangiophore. In slime molds, germination refers to the emergence of amoeboid cells from the hardened spore. After cracking the spore coat, further development involves cell division, but not necessarily the development of a multicellular organism (for example in the free–living amoebas of slime molds).

Ferns and Mosses

In plants such as bryophytes, ferns, and a few others, spores germinate into independent gametophytes. In the bryophytes (*e.g.*, mosses and liverworts), spores germinate into protonemata, similar to fungal hyphae, from which the gametophyte grows. In ferns, the gametophytes are small, heart–shaped prothalli that can often be found underneath a spore–shedding adult plant.

2

Bacterial Growth in Seed

INTRODUCTION

The factors discussed in this section constitute an inclusive, rather than exclusive, list of intrinsic, extrinsic, and other factors that may be considered when determining whether a food or category of foods requires time/temperature control during storage, distribution, sale and handling at retail and in food service to assure consumer protection.

Many factors must be evaluated for each specific food when making decisions on whether it needs time/temperature control for safety. These can be divided into intrinsic and extrinsic factors. Intrinsic factors are those that are characteristic of the food itself; extrinsic factors are those that refer to the environment surrounding the food. The need for time/temperature control is primarily determined by 1) the potential for contamination with pathogenic microorganisms of concern (including processing influences), and 2) the potential for subsequent growth and/or toxin production.

Most authorities are likely to divide foods among three categories based on an evaluation of the factors described below: those that do not need time/temperature control for protection of consumer safety; those that need time/temperature control; and those where the exact status is questionable. In the case of questionable products, further scientific evidence--such as modeling of microbial growth or death, or actual microbiological challenge studies--may help to inform the decision.

INTRINSIC FACTORS

MOISTURE CONTENT

Microorganisms need water in an available form to grow in food products. The control of the moisture content in foods is one of the oldest exploited preservation strategies. Food microbiologists generally describe the water requirements of microorganisms in terms of the water activity (aw) of the food or environment. Water activity is defined as the ratio of water vapor pressure of the food substrate to the vapor pressure of pure water at the same temperature:

$$aw = p/po,$$

where p = vapor pressure of the solution and po = vapor pressure of the solvent (usually water). The aw of pure water is 1.00 and the aw of a completely dehydrated food is 0.00. The aw of a food on this scale from 0.00 - 1.00 is related to the equilibrium relative humidity above the food on a scale of 0 - 100 per cent. Thus, per cent Equilibrium Relative Humidity (ERH) = aw x 100. The aw of a food describes the degree to which water is "bound" in the food, its availability to participate in chemical/biochemical reactions, and its availability to facilitate growth of microorganisms.

Most fresh foods, such as fresh meat, vegetables, and fruits, have awvalues that are close to the optimum growth level of most microorganisms (0.97 - 0.99). Table shows the approximate awlevels of some common food categories. The aw can be manipulated in foods by a number of means, including addition of solutes such as salt or sugar, physical removal of water through drying or baking, or binding of water to various macromolecular components in the food. Weight for weight, these food components will decrease awin the following order: ionic compounds > sugars, polyhydric alcohols, amino acids and other low-molecular-weight compounds > high-molecular-weight compounds such as cellulose, protein or starch.

Microorganisms respond differently to aw depending on a number of factors. Microbial growth, and, in some cases, the production of microbial metabolites, may be particularly sensitive to alterations in aw. Microorganisms generally have optimum and minimum levels of aw for growth depending on other growth factors in their environments. One indicator of microbial response is their taxonomic classification. For example, Gram (-) bacteria are generally more sensitive to low aw than Gram (+) bacteria. Table lists the approximate minimum aw values for the growth of selected microorganisms relevant to food. It should be noted that many bacterial pathogens are controlled at water activities well above 0.86 and only S. aureus can grow and produce toxin below aw 0.90. It must be emphasized that these are approximate values because solutes can vary in their ability to inhibit microorganisms at the same aw value. To illustrate, the lower aw limit for the growth of Clostridium botulinum type A has been found to be 0.94 with NaCl as the solute versus 0.92 with glycerol as the solute. When formulating foods using aw as the primary control mechanism for pathogens, it is useful to employ microbiological challenge testing to verify the effectiveness of the reduced aw when target aw is near the growth limit for the organism of concern.

Because aw limits vary with different solutes or humectants, other measures may provide more precise moisture monitoring for certain products. For example, factors other than aw are known to control the antibotulinal properties of pasteurized processed cheese spreads (Tanaka and others 1986). Also, aw may be used in combination with other factors to control pathogens in certain food products. Care should be taken when analyzing multicomponents foods, because effective measurements of aw may not reflect the actual value in a microenvironment or in the interface among the different components. In these cases, the awshould be measured at the interface areas of the food, as well as in any potential microenvironment.

Table. Approximate aw values of selected food categories.

Animal Products	aw
fresh meat, poultry, fish	0.99 - 1.00
natural cheeses	0.95 - 1.00
pudding	0.97 - 0.99
eggs	0.97
cured meat	0.87 - 0.95
sweetened condensed milk	0.83
Parmesan cheese	0.68 - 0.76
honey	0.75
dried whole egg	0.40
dried whole milk	0.20
Plant Products	aw
fresh fruits, vegetables	0.97 - 1.00
bread	~0.96
bread, white	0.94 - 0.97
bread, crust	0.30
baked cake	0.90 - 0.94
maple syrup	0.85
jam	0.75 - 0.80
jellies	0.82 - 0.94
uncooked rice	0.80 - 0.87
fruit juice concentrates	0.79 - 0.84
fruit cake	0.73 - 0.83
cake icing	0.76 - 0.84
flour	0.67 - 0.87
dried fruit	0.55 - 0.80
cereal	0.10 - 0.20
sugar	0.19
crackers	0.10

Table. Approximate aw values for growth of selected pathogens in food.

Organism	Minimum	Optimum	
Maximum			
Campylobacter spp.	0.98	0.99	
Clostridium botulinum type E*	0.97		
Shigella spp.	0.97		
Yersinia enterocolitica	0.97		
Vibrio vulnificus	0.96	0.98	0.99
Enterohemorrhagic Escherichia coli	0.95	0.99	

Salmonella spp.	0.94	0.99	>0.99
Vibrio parahaemolyticus	0.94	0.98	0.99
Bacillus cereus	0.93		
Clostridium botulinumtypes A & B**	0.93		
Clostridium perfringens	0.943	0.95-0.96	0.97
Listeria monocytogenes	0.92		
Staphylococcus aureus growth	0.83	0.98	0.99
Staphylococcus aureus toxin	0.88	0.98	0.99

PH AND ACIDITY

Increasing the acidity of foods, either through fermentation or the addition of weak acids, has been used as a preservation method since ancient times. In their natural state, most foods such as meat, fish, and vegetables are slightly acidic while most fruits are moderately acidic. A few foods such as egg white are alkaline. Table lists the pH ranges of some common foods. The pH is a function of the hydrogen ion concentration in the food:

$$pH = -\log10\ [H+]$$

Another useful term relevant to the pH of foods is the pKa. The term pKa describes the state of dissociation of an acid. At equilibrium, pKa is the pH at which the concentrations of dissociated and undissociated acid are equal. Strong acids have a very low pKa, meaning that they are almost entirely dissociated in solution. For example, the pH (at 25 °C [77 °F]) of a 0.1 M solution of HCl is 1.08 compared to the pH of 0.1 M solution of acetic acid, which is 2.6. This characteristic is extremely important when using acidity as a preservation method for foods. Organic acids are more effective as preservatives in the undissociated state. Lowering the pH of a food increases the effectiveness of an organic acid as a preservative. Table lists the proportion of total acid undissociated at different pH values for selected organic acids. The type of organic acid employed can dramatically influence the microbiological keeping quality and safety of the food.

It is well known that groups of microorganisms have pH optimum, minimum, and maximum for growth in foods. Table lists the approximate pH ranges for growth in laboratory media for selected organisms relevant to food. As with other factors, pH usually interacts with other parameters in the food to inhibit growth. The pH can interact with factors such as aw, salt, temperature, redox potential, and preservatives to inhibit growth of pathogens and other organisms. The pH of the food also significantly impacts the lethality of heat treatment of the food. Less heat is needed to inactivate microbes as the pH is reduced.

Another important characteristic of a food to consider when using acidity as a control mechanism is its buffering capacity. The buffering capacity of a food is its ability to resist changes in pH. Foods with a low buffering capacity will change pH quickly in response to acidic or alkaline compounds produced by microorganisms as they grow. Meats, in general, are more buffered than vegetables by virtue of their various proteins.

Titratable acidity (TA) is a better indicator of the microbiological stability of certain foods, such as salad dressings, than is pH. Titratable acidity is a measure of the quantity of standard alkali (usually 0.1 M NaOH) required to neutralize an acid solution. It measures the amount of hydrogen ions released from undissociated acid during titration. Titratable acidity is a particularly useful measure for highly buffered or highly acidic foods. Weak acids (such as organic acids) are usually undissociated and, therefore, do not directly contribute to pH. Titratable acidity yields a measure of the total acid concentration, while pH does not, for these types of foods.

In general, pathogens do not grow, or grow very slowly, at pH levels below 4.6; but there are exceptions. Many pathogens can survive in foods at pH levels below their growth minima. It has been reported that C. botulinumwas able to produce toxin as low as pH 4.2, but these experiments were conducted with high inoculum levels (103-104 CFU/g up to 106 CFU/g), in soy peptone, and with the presence of Bacillus spp. The panel did not consider these results to be relevant to the foods under consideration in this report. It should also be noted that changes in pH can transform a food into one which can support growth of pathogens. For example, several botulism outbreaks have been traced to foods in which the pH increased due to mold growth. These are important considerations when determining the shelf life of a food formulation. Based on a comprehensive review of the literature, the panel concluded that a pH of 4.6 is appropriate to control spore-forming pathogens.

Among vegetative pathogens, Salmonella spp. are reported to grow at the lowest pH values; however, in a study by Chung and Goepfert the limiting pH was greatly influenced by the acidulant used. For example, when tryptone-yeast extract-glucose broth was inoculated with 104 CFU/ml of salmonellae, minimum pH values for growth ranged from 4.05 with hydrochloric and citric acids to 5.5 with propionic acid or acetic acid. Additionally, inoculum levels were unrealistically high (102 - 106 CFU/ml) for salmonellae in food systems. These investigators also noted that these results could not be extrapolated directly to food because the experiment was run in laboratory media under ideal temperature and aw conditions and without the presence of competitive microorganisms. Similarly, Ferreira and Lund reported that six out of 13 strains of Salmonella spp. representing 12 serovars could grow at pH 3.8 at 30 °C (86 °F) within 1-3 d, and at 20 °C (68 °F) in 3-5 d, when using HCl as an acidulant. Other reports note that certain acids at pH 4.5 inactivate salmonellae. The panel therefore concluded that using a pH minimum of 4.0 for Salmonella spp. would not be scientifically substantiated for foods subject to Food Code requirements. Based on a comprenhensive review of the literature data, the panel also concluded that it would be scientifically valid to use a pH minimum of 4.2 to control for Salmonella spp. and other vegetative pathogens. As with other intrinsic properties, when analyzing multicomponents foods, the pH should be measured not only for each component of the food but also for the interface areas among components and for any potential microenvironment.

Table. pH ranges of some common foods.

Food	pH Range	
Dairy Products	Butter	6.1 - 6.4
	Buttermilk	4.5
	Milk	6.3 - 6.5
	Cream	6.5
	Cheese (American mild and cheddar)	4.9; 5.9
	Yogurt	3.8 - 4.2
Meat and Poultry	Beef (ground)	5.1 - 6.2
	Ham	5.9 - 6.1
	Veal	6.0
	Chicken	6.2 - 6.4
Fish and Shellfish	Fish (most species)	6.6 - 6.8
	Clams	6.5
	Crabs	7.0
	Oysters	4.8 - 6.3
	Tuna Fish	5.2 - 6.1
	Shrimp	6.8 - 7.0
	Salmon	6.1 - 6.3
	White Fish	5.5
Fruits and Vegetables	Apples	2.9 - 3.3
	Apple Cider	3.6 - 3.8
	Bananas	4.5 - 4.7
	Figs	4.6
	Grapefruit (juice)	3.0
	Limes	1.8 - 2.0
	Honeydew melons	6.3 - 6.7
	Oranges (juice)	3.6 - 4.3
	Plums	2.8 - 4.6
	Watermelons	5.2 - 5.6
	Grapes	3.4 - 4.5
	Asparagus (buds and stalks)	5.7 - 6.1
	Beans (string and lima)	4.6 - 6.5
	Beets (sugar)	4.2 - 4.4
	Broccoli	6.5
	Brussels Sprouts	6.3
	Cabbage (green)	5.4 - 6.0
	Carrots	4.9 - 5.2; 6.0
	Cauliflower	5.6
	Celery	5.7 - 6.0
	Corn (sweet)	7.3
	Cucumbers	3.8
	Eggplant	4.5
	Eggs yolks (white)	6.0 - 6.3 (7.6- 9.5)
	Lettuce	6.0

Olives (green)	3.6 - 3.8
Onions (red)	5.3 - 5.8
Parsley	5.7 - 6.0
Parsnip	5.3
Potatoes (tubers and sweet)	5.3 - 5.6
Pumpkin	4.8 - 5.2
Rhubarb	3.1 - 3.4
Spinach	5.5 - 6.0
Squash	5.0 - 5.4
Tomatoes (whole)	4.2 - 4.3
Turnips	5.2 - 5.5

Table. Proportion of total acid undissociated at different pH values (expressed as percentages).

Organic Acids	pH Values				
	3	4	5	6	7
Acetic acid	98.5	84.5	34.9	5.1	0.54
Benzoic acid	93.5	59.3	12.8	1.44	0.144
Citric acid	53.0	18.9	0.41	0.006	<0.001
Lactic acid	86.6	39.2	6.05	0.64	0.064
Methyl, ethyl, propyl parabens	>99.99	99.99	99.96	99.66	96.72
Propionic acid	98.5	87.6	41.7	6.67	0.71
Sorbic acid	97.4	82.0	30.0	4.1	0.48

Table. Approximate pH values permitting the growth of selected pathogens in food.

Microorganism	Minimum	Optimum	Maximum
Clostridium perfringens	5.5 - 5.8	7.2	8.0 - 9.0
Vibrio vulnificus	5.0	7.8	10.2
Bacillus cereus	4.9	6.0 -7.0	8.8
Campylobacter spp.	4.9	6.5 - 7.5	9.0
Shigella spp.	4.9		9.3
Vibrio parahaemolyticus	4.8	7.8 - 8.6	11.0
Clostridium botulinum toxin	4.6		8.5
Clostridium botulinum growth	4.6		8.5
Staphylococcus aureus growth	4.0	6.0 - 7.0	10.0
Staphylococcus aureus toxin	4.5	7.0 - 8.0	9.6
Enterohemorrhagic Escherichia coli	4.4	6.0 - 7.0	9.0
Listeria monocytogenes	4.39	7.0	9.4
Salmonella spp.	4.21	7.0 - 7.5	9.5
Yersinia enterocolitica	4.2	7.2	9.6

NUTRIENT CONTENT

Microorganisms require certain basic nutrients for growth and maintenance of metabolic functions. The amount and type of nutrients required range widely depending on the microorganism. These nutrients include water, a source of energy, nitrogen, vitamins, and minerals. Varying amounts of these nutrients are present in foods. Meats have abundant protein, lipids, minerals, and vitamins.

Most muscle foods have low levels of carbohydrates. Plant foods have high concentrations of different types of carbohydrates and varying levels of proteins, minerals, and vitamins. Foods such as milk and milk products and eggs are rich in nutrients.

Foodborne microorganisms can derive energy from carbohydrates, alcohols, and amino acids. Most microorganisms will metabolize simple sugars such as glucose. Others can metabolize more complex carbohydrates, such as starch or cellulose found in plant foods, or glycogen found in muscle foods. Some microorganisms can use fats as an energy source.

Amino acids serve as a source of nitrogen and energy and are utilized by most microorganisms. Some microorganisms are able to metabolize peptides and more complex proteins. Other sources of nitrogen include, for example, urea, ammonia, creatinine, and methylamines.

Examples of minerals required for microbial growth include phosphorus, iron, magnesium, sulfur, manganese, calcium, and potassium. In general, small amounts of these minerals are required; thus a wide range of foods can serve as good sources of minerals.

In general, the Gram (+) bacteria are more fastidious in their nutritional requirements and thus are not able to synthesize certain nutrients required for growth. For example, the Gram (+) foodborne pathogen S. aureus requires amino acids, thiamine, and nicotinic acid for growth. Fruits and vegetables that are deficient in B vitamins do not effectively support the growth of these microorganisms. The Gram (-) bacteria are generally able to derive their basic nutritional requirements from the existing carbohydrates, proteins, lipids, minerals, and vitamins that are found in a wide range of food.

An example of a pathogen with specific nutrient requirements is Salmonella Enteritidis. Growth of SalmonellaEnteritidis may be limited by the availability of iron. For example, the albumen portion of the egg, as opposed to the yolk, includes antimicrobial agents and limited free iron that prevent the growth of Salmonella Enteritidis to high levels. Clay and Board demonstrated that the addition of iron to an inoculum of Salmonella Enteritidis in egg albumen resulted in growth of the pathogen to higher levels compared to levels reached when a control inoculum (without iron) was used.

The microorganisms that usually predominate in foods are those that can most easily utilize the nutrients present. Generally, the simple carbohydrates and amino acids are utilized first, followed by the more complex forms of these nutrients. The complexity of foods in general is such that several microorganisms can be growing in a food at the same time. The rate of growth is limited by the availability of essential nutrients. The abundance of nutrients in most foods is sufficient to support the growth of a wide range of foodborne pathogens. Thus, it is very difficult and impractical to predict the pathogen growth or toxin production based on the nutrient composition of the food.

BIOLOGICAL STRUCTURE

Plant and animal derived foods, especially in the raw state, have biological structures that may prevent the entry and growth of pathogenic microorganisms. Examples of such physical barriers include testa of seeds, skin of fruits and vegetables, shell of nuts, animal hide, egg cuticle, shell, and membranes.

Plant and animal foods may have pathogenic microorganisms attached to the surface or trapped within surface folds or crevices. Intact biological structures thus can be important in preventing entry and subsequent growth of microorganisms. Several factors may influence penetration of these barriers. The maturity of plant foods will influence the effectiveness of the protective barriers. Physical damage due to handling during harvest, transport, or storage, as well as invasion of insects can allow the penetration of microorganisms. During the preparation of foods, processes such as slicing, chopping, grinding, and shucking will destroy the physical barriers. Thus, the interior of the food can become contaminated and growth can occur depending on the intrinsic properties of the food. For example,Salmonella spp. have been shown to grow on the interior of portions of cut cantaloupe, watermelon, honeydew melons, and tomatoes, given sufficient time and temperature.

Fruits are an example of the potential of pathogenic microorganisms to penetrate intact barriers. After harvest, pathogens will survive but usually not grow on the outer surface of fresh fruits and vegetables. Growth on intact surfaces is not common because foodborne pathogens do not produce the enzymes necessary to break down the protective outer barriers on most produce.

This outer barrier restricts the availability of nutrients and moisture. One exception is the reported growth of E. coli O157:H7 on the surface of watermelon and cantaloupe rinds. Survival of foodborne pathogens on produce is significantly enhanced once the protective epidermal barrier has been broken either by physical damage, such as punctures or bruising, or by degradation by plant pathogens (bacteria or fungi). These conditions can also promote the multiplication of pathogens, especially at higher temperatures. Infiltration of fruit was predicted and described by Bartz and Showalter based on the general gas law, which states that any change in pressure of an ideal gas in a closed container of constant volume is directly proportional to a change in temperature of the gas.

In their work, Bartz and Showalter describe a tomato; however, any fruit, such as an apple, can be considered a container that is not completely closed. As the container or fruit cools, the decrease in internal gas pressure results in a partial vacuum inside the fruit, which then results in an influx from the external environment. For example, an influx of pathogens from the fruit surface or cooling water could occur as a result of an increase in external pressure due to immersing warm fruit in cool water. Internalization of bacteria into fruits and vegetables could also occur due to breaks in the tissues or through morphological structures in the fruit itself, such as the calyx or stem scar. Although infiltration was considered a possible scenario, the panel concluded that there is insufficient epidemiological evidence to require refrigeration of intact fruit.

The egg is another good example of an effective biological structure that, when intact, will prevent external microbial contamination of the perishable yolk; contamination is possible, however, through transovarian infection. For the interior of an egg to become contaminated by microorganisms on the surface, there must be penetration of the shell and its membranes. In addition, the egg white contains antimicrobial factors. When there are cracks through the inner membrane of the egg, microorganisms penetrate into the egg. Factors such as temperature of storage, relative humidity, age of eggs, and level of surface contamination will influence internalization. For example, conditions such as high humidity and wet and dirty shells, along with a drop in the storage temperature will increase the likelihood for entry of bacteria. If eggs are washed, the wash water should be 12 °C (22 °F) higher than the temperature of the eggs to prevent microbial penetration. After washing, the eggs should be dried and then cooled. The Food and Drug Administration (FDA) published a final rule that applies to shell eggs that have not be processed to destroy all live Salmonella before distribution to the consumer. The rule mandates that eggs should be kept dry and chilled below 7.2 °C (45 °F) to prevent growth of Salmonella Enteritidis.

Heating of food as well as other types of processing will break down protective biological structures and alter such factors as pH and aw. These changes could potentially allow the growth of microbial pathogens.

REDOX POTENTIAL

The oxidation-reduction or redox potential of a substance is defined in terms of the ratio of the total oxidizing (electron accepting) power to the total reducing (electron donating) power of the substance. In effect, redox potential is a measurement of the ease by which a substance gains or loses electrons. The redox potential (Eh) is measured in terms of millivolts. A fully oxidized standard oxygen electrode will have an Eh of +810 mV at pH 7.0, 30 °C (86 °F), and under the same conditions, a completely reduced standard hydrogen electrode will have an Eh of -420 mV. The Eh is dependent on the pH of the substrate; normally the Eh is taken at pH 7.0.

The major groups of microorganisms based on their relationship to Eh for growth are aerobes, anaerobes, facultative aerobes, and microaerophiles. Examples of foodborne pathogens for each of these classifications include Aeromonas hydrophila, Clostridium botulinum, Escherichia coli O157:H7, and Campylobacter jejuni,respectively. Generally, the range at which different microorganisms can grow are as follows: aerobes +500 to +300 mV; facultative anaerobes +300 to -100 mV; and anaerobes +100 to less than -250 mV. For example, C. botulinum is a strict anaerobe that requires an Eh of less than +60 mV for growth; however, slower growth can occur at higher Eh values. The relationship of Eh to growth can be significantly affected by the presence of salt and other food constituents. For example, in one study with smoked herring, toxin was produced in inoculated product stored at 15 °C (59 °F) within three days at an Eh of +200 to +250 mV (Huss and others 1979). In this case, the major oxidant would be trimethylamine oxide, which becomes the electron acceptor for C. botulinum. The anaerobe Clostridium perfringens can initiate growth at an Eh close to +200 mV; however, in the presence of increasing concentrations of certain substances, such as salt, the limiting Eh increases.

These values can be highly variable depending on changes in the pH of the food, microbial growth, packaging, the partial pressure of oxygen in the storage environment, and ingredients and composition (protein, ascorbic acid, reducing sugars, oxidation level of cations, and so on). Another important factor is the poising capacity of the food. Poising capacity, which is analogous to buffering capacity, relates to the extent to which a food resists external affected changes in Eh. The poising capacity of the food will be affected by oxidizing and reducing constituents in the food as well as by the presence of active respiratory enzyme systems. Fresh fruits and vegetables and muscle foods will continue to respire; thus low Eh values can result.

The measurement of redox potential of food is done rather easily, either for single or multicomponent foods. For multicomponent foods, in addition to measurement of each component, the redox potential of the interface areas and microenvironments should be considered. However, difficulties arise in taking accurate measurements and in accounting for the differences throughout the food and the equilibrium at the point of measurement. According to Morris: "This imposes the further requirements 1) that the measuring electrode be so prepared and calibrated that it gives stable and reproducible readings, and 2) that a foodstuff is tested in a manner that does not cause any change in the potential that is to be measured.... it would be unwise to use redox potential information in isolation to predict food safety, or to rely exclusively on control of redox potential as the means of preventing growth of specific microorganisms." Redox measurements could possibly be used in combination with other factors to evaluate the potential for pathogen growth. However, the limitations discussed above make it a rather difficult and variable factor that could result in erroneous conclusions in the absence of other comprehensive information.

NATURALLY OCCURRING AND ADDED ANTIMICROBIALS

Some foods intrinsically contain naturally-occurring antimicrobial compounds that convey some level of microbiological stability to them. There are a number of plant-based antimicrobial constituents, including many essential oils, tannins, glycosides, and resins, that can be found in certain foods. Specific examples include eugenol in cloves, allicin in garlic, cinnamic aldehyde and eugenol in cinnamon, allyl isothiocyanate in mustard, eugenol and thymol in sage, and carvacrol (isothymol) and thymol in oregano. Other plant-derived antimicrobial constituents include the phytoalexins and the lectins. Lectins are proteins that can specifically bind to a variety of polysaccharides, including the glycoproteins of cell surfaces (Mossel and others 1995, p 175-214). Through this binding, lectins can exert a slight antimicrobial effect. The usual concentration of these compounds in formulated foods is relatively low, so that the antimicrobial effect alone is slight. However, these compounds may produce greater stability in combination with other factors in the formulation.

Some animal-based foods also contain antimicrobial constituents. Examples include lactoferrin, conglutinin and the lactoperoxidase system in cow's milk, lysozyme in eggs and milk, and other factors in fresh meat, poultry and seafood. Lysozyme is a small protein that can hydrolyze the cell wall of bacteria. The lactoperoxidase system in bovine milk consists of three distinct components that are required for its antimicrobial action: lactoperoxidase, thiocyanate, and hydrogen peroxide. Gram (-) psychotrophs such as the pseudomonads have been shown to be very sensitive to the lactoperoxidase system. Consequently, this system, in an enhanced form, has been suggested to improve the keeping quality of raw milk in developing countries where adequate refrigeration is scarce. Similar to the plant-derived antimicrobial compounds, the animal-derived compounds have a limited effect on ambient shelf life of foods.

It is also known that some types of food processing result in the formation of antimicrobial compounds in the food. The smoking of fish and meat can result in the deposition of antimicrobial substances onto the product surface. Maillard compounds resulting from condensation reactions between sugars and amino acids or peptides upon heating of certain foods can impart some antimicrobial activity. Smoke condensate includes phenol, which is not only an antimicrobial, but also lowers the surface pH. Some procesors also lower the surface pH with liquid smoke to achieve an unsliced shelf-stable product.

Some types of fermentations can result in the natural production of antimicrobial substances, including bacteriocins, antibiotics, and other related inhibitors. Bacteriocins are proteins or peptides that are produced by certain strains of bacteria that inactivate other, usually closely-related, bacteria. The most commonly characterized bacteriocins are those produced by the lactic acid bacteria. The lantibiotic nisin produced by certain strains of Lactococcus lactis is one of the best characterized of the bacteriocins. Nisin is approved for food applications in over 50 countries around the world (Jay 2000, p 269-72). Nisin's first food application was to prevent late-blowing in Swiss cheese by

Clostridium butyricum. Nisin is a polypeptide that is effective against most Gram (+) bacteria but is ineffective against Gram (-) organisms and fungi. Nisin can be produced in the food by starter cultures or, more commonly, it can be used as an additive in the form of a standardized preparation (Lück and Jager 1997). Nisin has been used to effectively control spore-forming organisms in processed cheese formulations, and has been shown to have an interactive effect with heat. For example, an Fo process for conventional low acid canned foods may be in the 6 - 8 range, but with the addition of nisin, can be reduced to a Fo of 3 for inactivating thermophilic spores.

There are a number of other bacteriocins and natural antimicrobials that have been described, however, these have found very limited application in commercial use as food preservatives because of their restricted range of activity, limited compatibility with the food formulation or their regulatory status.

In addition to naturally-occurring antimicrobial compounds in foods, a variety of chemical preservatives and additives can extend the shelf life of food and/or inhibit pathogens, either singly or in combination. The selection and use of these preservatives is typically governed by food law regulation of a country or region of the world. A number of criteria should be followed when selecting a preservative for a specific food application. Ideally, the preservative should have a wide spectrum of activity against the target spoilage organisms and pathogens expected to be encountered in the food. The preservative must be active for the desired shelf life of the food and under the expected formulation conditions in the food. It should cause minimal organoleptic impact on the food and should not interfere with desirable microbiological processes expected to occur in the food, such as the ripening of cheese or leavening of baked goods.

Added antimicrobial compounds can have an interactive or synergistic effect with other parameters of the formulation. One example is the interaction with pH. Many preservatives have an optimum pH range for effectiveness. Other factors include aw, presence of other preservatives, types of food constituents, presence of certain enzymes, processing temperature, storage atmosphere, and partition coefficients. The effective use of combinations of preservatives with other physico-chemical parameters of a food formulation can stabilize that food against spoilage organisms or pathogens. Leistner systematically developed the "hurdle concept" to describe these effects. The hurdle concept states that several inhibitory factors (hurdles), while individually unable to inhibit microorganisms, will, nevertheless, be effective in combination. A classic example of applying the hurdle concept is the anti-botulinal stability of certain shelf-stable processed cheese formulations. Combinations of moisture, total salt, and pH have been shown to allow for the safe storage of these products at room temperature for extended time even though the individual factors, taken singly, would not support that practice. In combination products, the effectiveness of an antimicrobial may be altered by other factors including the potential for migration of the antimicrobial to other components of the food and the different food parameters at the interface areas.

There are a number of food formulations that, either by addition of preservatives or through the application of the hurdle concept do not require refrigeration for microbiological stability or safety. However, in the absence of a well-defined and validated microbiological model, it is usually difficult to evaluate the microbiological safety of these products. In the majority of these cases, the application of appropriate microbiological challenge testing is the most effective tool for judging the suitability of these formulations for non-refrigerated storage.

COMPETITIVE MICROFLORA

The potential for microbial growth of pathogens in temperature-sensitive foods depends on the combination of the intrinsic and extrinsic factors, and the processing technologies that have been applied. Within the microbial flora in a food, there are many important biological attributes of individual organisms that influence the species that predominates. These include the individual growth rates of the microbial strains and the mutual interactions or influences among species in mixed populations.

Growth

In a food environment, an organism grows in a characteristic manner and at a characteristic rate. The length of the lag phase, generation time, and total cell yield are determined by genetic factors. Accumulation of metabolic products may limit the growth of particular species. If the limiting metabolic product can be used as a substrate by other species, these may take over (partly or wholly), creating an association or succession. Due to the complex of continuing interactions between environmental factors and microorganisms, a food at any one point in time has a characteristic flora, known as its association. The microbial profile changes continuously and one association succeeds another in what is called succession. Many examples of this phenomenon have been observed in the microbial deterioration and spoilage of foods.

As long as metabolically active organisms remain, they continue to interact, so that dominance in the flora occurs as a dynamic process. Based on their growth-enhancing or inhibiting nature, these interactions are either antagonistic or synergistic.

Competition

In food systems, antagonistic processes usually include competition for nutrients, competition for attachment/adhesion sites (space), unfavorable alterations of the environment, and a combination of these factors. Early studies demonstrated that the natural biota of frozen pot pies inhibited inoculated cells of S. aureus, E. coli and Salmonella Typhimurium. Another example of this phenomenon is raw ground beef. Even though S. aureus is often found in low numbers in this product, staphylococcal enterotoxin is not produced. The reason is that the Pseudomonas-Acinetobacter-Moraxella association that is always present in this food grows at a higher rate, outgrowing the staphylococci. Organisms of high metabolic activity may consume required nutrients, selectively reducing these substances, and inhibiting the growth of other organisms. Depletion of oxygen or accumulation of carbon dioxide favours facultative obligate anaerobes which occur in vacuum-packaged fresh meats held under refrigeration.

Staphylococci are particularly sensitive to nutrient depletion. Coliforms and Pseudomonas spp. may utilize amino acids necessary for staphylococcal growth and make them unavailable. Other genera of Micrococcaceae can utilize nutrients more rapidly than staphylococci. Streptococci inhibit staphylocci by exhausting the supply of nicotinamide or niacin and biotin. Staphylococcus aureus is a poor competitor in both fresh and frozen foods. At temperatures that favour staphylococcal growth, the normal food saprophytic biota offers protection against staphylococcal growth through antagonism, competition for nutrients, and modification of the environment to conditions less favorable to S. aureus. Changes in the composition of the food, as well as changes in intrinsic or extrinsic factors may either stimulate or decrease competitive effects.

Effects on growth inhibition

Changes in growth stimulation have been reported among several foodborne organisms, including yeasts, micrococci, streptococci, lactobacilli and Enterobacteriaceae. Growth stimulating mechanisms can have a significant influence on the buildup of a typical flora. There are several of these mechanisms, a few of which are listed below:

- Metabolic products from one organism can be absorbed and utilized by other organisms.
- Changes in pH may promote the growth of certain microorganisms. An example is natural fermentations, in which acid production establishes the dominance of acid tolerant organisms such as the lactic acid bacteria. Growth of molds on high acid foods has been found to raise the pH, thus stimulating the growth ofC. botulinum.
- Changes in Eh or aw in the food can influence symbiosis. At warm temperatures, C. perfringens can lower the redox potential in the tissues of freshly slaughtered animals so that even more obligately anaerobic organisms can grow.
- There are some associations where maximum growth and normal metabolic activity are not developed unless both organisms are present.

This information can be used in the hurdle concept to control microorganisms in temperature-sensitive foods.

EXTRINSIC FACTORS

TYPES OF PACKAGING/ATMOSPHERES

Many scientific studies have demonstrated the antimicrobial activity of gases at ambient and sub-ambient pressures on microorganisms important in foods.

Gases inhibit microorganisms by two mechanisms. First, they can have a direct toxic effect that can inhibit growth and proliferation. Carbon dioxide (CO_2), ozone (03), and oxygen (O_2) are gases that are directly toxic to certain microorganisms. This inhibitory mechanism is dependent upon the chemical and physical properties of the gas and its interaction with the aqueous and lipid phases of the food. Oxidizing radicals generated by 03and O_2 are highly toxic to anaerobic bacteria and can have an inhibitory effect on aerobes depending on their concentration. Carbon dioxide is effective against obligate aerobes and at high levels can deter other microorganisms. A second inhibitory mechanism is achieved by modifying the gas composition, which has indirect inhibitory effects by altering the ecology of the microbial environment. When the atmosphere is altered, the competitive environment is also altered. Atmospheres that have a negative effect on the growth of one particular microorganism may promote the growth of another. This effect may have positive or negative consequences depending upon the native pathogenic microflora and their substrate. Nitrogen replacement of oxygen is an example of this indirect antimicrobial activity.

A variety of common technologies are used to inhibit the growth of microorganisms, and a majority of these methods rely upon temperature to augment the inhibitory effects. Technologies include modified atmosphere packing (MAP), controlled atmosphere packaging (CAP), controlled atmosphere storage (CAS), direct addition of carbon dioxide (DAC), and hypobaric storage.

Controlled atmosphere and modified atmosphere packaging of certain foods can dramatically extend their shelf life. The use of CO_2, N_2, and ethanol are examples of MAP applications. In general, the inhibitory effects of CO_2 increase with decreasing temperature due to the increased solubility of CO_2 at lower temperatures. Carbon dioxide dissolves in the food and lowers the pH of the food. Nitrogen, being an inert gas, has no direct antimicrobial properties. It is typically used to displace oxygen in the food package either alone or in combination with CO_2, thus having an indirect inhibitory effect on aerobic microorganisms. Table shows some examples of combinations of gases for MAP applications in meat, poultry, seafood, hard cheeses, and baked goods.

Table. Examples of gas mixtures used for various MAP products.

Product	% CO_2	% O_2	% N_2
Fresh meat	30	30	40
	15 - 40	60 - 85	0
Cured meat	20 - 50	0	50 - 80
Sliced cooked roast beef	75	10	15
Eggs	20	0	80
	0	0	100
Poultry	25 - 30	0	70 - 75
	60 - 75	5 - 10	>e 20
	100	0	0
	20-40	60-80	0
Pork	20	80	0
Processed Meats	0	0	100
Fish (White)	40	30	30
Fish (Oily)	40	0	60
	60	0	40
Hard cheese	0 - 70		30 - 100
Cheese	0	0	100
Cheese; grated/sliced	30	0	70
Sandwiches	20 - 100	0 - 10	0 - 100
Pasta	0	0	100
	70 - 80	0	20 - 30
Baked goods	20 - 70	0	20 - 80
	0	0	100
	100	0	0

The preservation principle of antimicrobial atmospheres has been applied to fruits and vegetables, raw beef, chicken and fish, dairy foods including milk and cottage cheese, eggs, and a variety of prepared, ready-to-eat foods.

There are several intrinsic and extrinsic factors that influence the efficacy of antimicrobial atmospheres. These factors-including product temperature, product-to-headspace gas volume ratio, initial microbial loads and type of flora, package barrier properties, and biochemical composition of the food-all interact to determine the degree to which the microbial quality and safety are enhanced.

Temperature, the most important factor affecting the efficacy of antimicrobial atmospheres, directly affects growth rate, but also indirectly affects growth by affecting gas solubility. At practical food storage temperatures, packaging configurations, especially the product-to-headspace volume ratio, play a major role in determining the magnitude of microbial inhibition.

In MAP, package barrier properties have a major effect on the microbial growth by influencing the time in which the selected modified atmosphere gases remain in contact with the product and the rate at which oxygen enters the package.

Water activity, salt content of the aqueous phase, pH, and fat content of foods also play a role in overall inhibitory effects of antimicrobial gases. As with temperature, the physical and chemical characteristics of the food have an effect on the solubility of the inhibitory gas. For example, increasing salt concentrations decreases CO_2 solubility.

The major safety consideration in extending shelf life of foods by MAP or related technologies is the loss of sensory cues to spoilage provided by bacterial growth. Without spoilage bacteria indicators, it is conceivable that a food could have acceptable organoleptic quality, but be unsafe. The effect of loss of competitive inhibition by spoilage bacteria is most pronounced on the facultative anaerobic pathogenic bacterial populations in foods under altered atmospheres.

By combining antimicrobial atmospheres with other techniques, hurdle technology strategies may be generated that can further enhance food quality and safety.

EFFECT OF TIME/TEMPERATURE CONDITIONS ON MICROBIAL GROWTH

Impact of time

When considering growth rates of microbial pathogens, in addition to temperature, time is a critical consideration. Food producers or manufacturers address the concept of time as it relates to microbial growth when a product's shelf life is determined. Shelf life is the time period from when the product is produced until the time it is intended to be consumed or used. Several factors are used to determine a product's shelf life, ranging from organoleptic qualities to microbiological safety. For the purpose of this report, the key consideration is the microbiological safety of the product. The Uniform Open Dating Regulation requires the shelf life of a perishable food product to be expressed in terms of a "sell by" date. The "sell by" date must incorporate the shelf life of the product plus a reasonable period for consumption that consists of at least one-third of the approximate total shelf life of the perishable food product.

At retail or foodservice, an additional period of time referred to herein as "use-period" should also be considered. As an example, fast food locations may find it operationally desirable to hold processed cheese slices at ambient temperatures for a complete shift or meal period, which may be in excess of 4 h. This practice provides operational efficiency by allowing the cheese to melt faster on a hot sandwich as well as providing a better quality sandwich. Although refrigeration may be required for safety under long-term storage conditions, for use-periods measured in hours, storage at ambient temperatures may be acceptable.

Under certain circumstances, time alone at ambient temperatures can be used to control product safety. When time alone is used as a control, the duration should be equal to or less than the lag phase of the pathogen(s) of concern in the product in question. For refrigerated food products, the shelf life or use-period required for safety may vary depending on the temperature at which the product is stored. For example, Mossel and Thomas (1988) report that the lag time for growth of L. monocytogenes at 10 °C (50 °F) is 1.5 d, while at 1 °C (34 °F) lag time is ~3.3 d. Likewise, they report that at 10 °C (50 °F) the generation time for the same organism is 5-8 h, while at 1 °C (34 °F), the generation time is between 62 and 131 h.

The data were obtained by using the USDA Pathogen Micromodel Programme at a NaCl concentration of 2 per cent and aw of 0.989. It should be noted that this model was developed in broth under various salt and pH combinations, and that growth of bacteria in food systems will likely differ. According to the model results, a temperature shift from 10 (50) to 25 °C (77 °F) decreases the lag time of L. monocytogenes from 60 to 10 h. In a similar manner, a pH increase from 4.5 to 6.5 decreases the lag time from 60 to 5 h. In conclusion, the safety of a product during its shelf life may differ, depending upon other conditions such as temperature of storage, pH of the product, and so on. This study by Mossel and Thomas, along with numerous others, illustrates that various time/temperature combinations can be used to control product safety depending on the product's intended use.

Effect of temperature or pH on lag times of Listeria monocytogenes from USDA PMP ver 5.1 (2 per cent NaCl, aw 0.989)

As stated earlier, time alone at ambient temperatures can be used to control product safety. When time alone is used as a control, the duration should be equal to or less than the lag phase of the pathogen(s) of concern in the product in question.

Impact of temperature

All microorganisms have a defined temperature range in which they grow, with a minimum, maximum, and optimum. An understanding of the interplay between time, temperature, and other intrinsic and extrinsic factors is crucial to selecting the proper storage conditions for a food product. Temperature has dramatic impact on both the generation time of an organism and its lag period. Over a defined temperature range, the growth rate of an organism is classically defined as an Arrhenius relationship. The log growth rate constant is found to be proportional to the reciprocal of the absolute temperature:

$G = -\mu / 2.303\ RT$ where,

G = log growth rate constant

μ = temperature characteristic (constant for a particular microbe)

R = gas constant

T = temperature (°K)

The above relationship holds over the linear portion of the Arrhenius plot. However, when temperatures approach the maxima for a specific microorganism, the growth rate declines more rapidly than when temperatures approach the minima for that same microorganism. A relationship that more accurately predicts growth rates of microorganisms at low temperatures follows:

vr = b(T - To) where,

r = growth rate

b = slope of the regression line

T = temperature (°K)

To = conceptual temperature of no metabolic significance

At low temperatures, two factors govern the point at which growth stops: 1) reaction rates for the individual enzymes in the organism become much slower, and 2) low temperatures reduce the fluidity of the cytoplasmic membrane, thus interfering with transport mechanisms. At high temperatures, structural cell components become denatured and inactivation of heat-sensitive enzymes occurs. While the growth rate increases with increasing temperature, the rate tends to decline rapidly thereafter, until the temperature maximum is reached.

The relationship between temperature and growth rate constant varies significantly across groups of microorganisms. Four major groups of microorganisms have been described based on their temperature ranges for growth: thermophiles, mesophiles, psychrophiles, and psychrotrophs. Tables list the temperature ranges for these four groups (ICMSF 1980) and for pathogens of concern. The optimum temperature for growth of thermophiles is between 55 to 65 °C (131 to 149 °F) with the maximum as high as 90 °C (194 °F) and a minimum of around 40 °C (104 °F). Mesophiles, which include virtually all human pathogens, have an optimum growth range of between 30 °C (86 °F) and 45 °C (113 °F), and a minimum growth temperature ranging from 5 to 10 °C (41 to 50 °F). Psychrophilic organisms have an optimum growth range of 12 °C (54 °F) to 15 °C (59 °F) with a maximum range of 15 °C (59 °F) to 20 °C (68 °F). There are very few true psychrophilic organisms of consequence to foods. Psychrotrophs such as L. monocytogenes and C. botulinum type E are capable of growing at low temperatures (minimum of - 0.4 °C [31 °F] and 3.3 °C [38 °F], respectively, to 5 °C [41 °F]), but have a higher growth optimum range (37 °C [99 °F] and 30 °C [86 °F], respectively) than true psychrophiles. Psychrotrophic organisms are much more relevant to food and include spoilage bacteria, spoilage yeast and molds, as well as certain foodborne pathogens.

Growth temperature is known to regulate the expression of virulence genes in certain foodborne pathogens. For example, the expression of proteins governed by the Yersinia enterocoliticavirulence plasmid is high at 37 °C (99 °F), low at 22 °C (72 °F), and not detectable at 4 °C (39 °F). Growth temperature also impacts an organism's thermal sensitivity. Listeria monocytogenes, when held at 48 °C (118 °F) in inoculated sausages, has an increase of 2.4-fold in its D value at 64 °C (147 °F). It must be emphasized that the lag period and growth rate of a microorganism are influenced not only by temperature but by other intrinsic and extrinsic factors as well. For example, as shown in Table, the growth rate of Clostridium perfringens is significantly lower at pH 5.8 versus pH 7.2 across a wide range of temperatures. Salmonellae do not grow at temperatures below 5.2 °C (41 °F). The intrinsic factors of the food product, however, have been shown to impact the ability of salmonellae to grow at low temperatures. Salmonella Senftenberg, S. Enteritidis, and S. Manhattan were not able to grow in ham salad or custard held at 10 °C (50 °F), but were able to grow in chicken à la king held at 7 °C (45 °F) (ICMSF 1980, p 9). Staphylococcus aureus has been shown to grow at temperatures as low as 7 °C (45 °F), but the lower limit for enterotoxin production has been shown to be 10 °C (50 °F). In general, toxin production below about 20 °C (68 °F) is slow. For example, in laboratory media at pH 7, the time to produce detectable levels of enterotoxin ranged from 78 - 98 h at 19 °C (66 °F) to 14 - 16 h at 26 °C (79 °F). Less favorable conditions, such as reduced pH, slowed enterotoxin production even further.

Table illustrates the combined impact of temperature, pH, and aw on the growth of proteolytic C. botulinum type B. This table clearly shows that an interactive effect occurs between these three factors. When measuring the suitability of holding a refrigerated food at room temperature for a period of time, consideration may be given to each factor independently. Doing so, however, ignores the potential to safely hold products for a period of time out of refrigeration based on interaction effects. Consideration of each relevant factor independently may lead to the conclusion that it is not a safe practice to do so, while, in reality, it is actually safe based on the interactive effects. The most appropriate method for evaluating such interactive effects is through a properly designed microbiological challenge study using relevant target microorganisms. Appropriate, validated predictive microbiological models may also be employed for this purpose. The use of challenge studies and/or predictive models can yield scientific data that supports holding a product with a certain formulation for a given time and temperature. It is incumbent upon the producer to have specific knowledge of the food formulation to generate valid scientific data.

Table. Temperature ranges for prokaryotic microorganisms.

Group	Temperature °C (°F)		
	Minimum	Optimum	Maximum
Thermophiles	40 - 45 (104 - 113)	55 - 75 (131 - 167)	60 - 90 (140 - 194)
Mesophiles	5 - 15 (41 - 59)	30 - 45 (86 - 113)	35 - 47 (95 - 117)
Psychrophiles	-5 - +5 (23 - 41)	12 - 15 (54 - 59)	15 - 20 (59 - 68)
Psychrotrophs	-5 - +5 (23 - 41)	25 - 30 (77 - 86)	30 - 35 (86 - 95)

Table. Approximate minimum, maximum and optimum temperature values in °C (°F) permitting growth of selected pathogens relevant to food.

Organism	Minimum	Optimum	Maximum
Bacillus cereus	5 (41)	28 - 40 (82 - 104)	55 (131)
Campylobacter spp.	32 (90)	42 - 45 (108 - 113)	45 (113)

Clostridium botulinum types A & B*	10 - 12 (50 - 54)	30 - 40 (86 - 104)	50 (122)
Clostridium botulinum type E**	3 - 3.3 (37 - 38)	25 - 37 (77 - 99)	45 (113)
Clostridium perfringens	12 (54)	43 - 47 (109 - 117)	50 (122)
Enterotoxigenic Escherichia coli	7 (45)	35 - 40 (95 - 104)	46 (115)
Listeria monocytogenes	0 (32)	30 - 37 (86 - 99)	45 (113)
Salmonella spp.	5 (41)	35 - 37 (95 - 99)	45 - 47 (113 - 117)
Staphylococcus aureus growth	7 (45)	35 - 40 (95 - 104)	48 (118)
Staphylococcus aureus toxin	10 (50)	40 - 45 (104 - 113)	46 (115)
Shigella spp.	7 (45)	37 (99)	45 - 47 (113 - 117)
Vibrio cholerae	10 (50)	37 (99)	43 (109)
Vibrio parahaemolyticus	5 (41)	37 (99)	43 (109)
Vibrio vulnificus	8 (46)	37 (99)	43 (109)
Yersinia enterocolitica	-1 (30)	28 - 30 (82 - 86)	42 (108)

Table. The relationship of pH and temperature to growth rate of Clostridium perfringens(welchii) F2985/50.

Incubation temperature	Hours to visible turbidity in RCM broth at pH	
	5.8	7.2
15 °C (59 °F)	>700	>700
20 °C (68 °F)	74	48
25 °C (77 °F)	30	24
30 °C (86 °F)	24	8
37 °C (99 °F)	5	5

Table. Incubation period, in days, before growth of proteolytic Clostridium botulinum type B was observed at various levels of temperature, pH, and aw.

Temperature	pH	aw						
		0.997	0.99	0.98	0.97	0.96	0.95	0.94
20 °C (68 °F)	5	--	--	--	--	--	--	--
	6	49	9	--	--	--	--	--
	7	2	2	4	9	--	--	--
	8	2	2	4	14	--	--	--
	9	--	--	--	--	--	--	--
30 °C (86 °F)	5	--	--	--	--	--	--	--
	6	2	2	3	9	--	--	--
	7	1	1	2	3	9	14	--
	8	1	1	2	4	14	--	--
	9	--	--	--	--	--	--	--
40 °C (104 °F)	5	--	--	--	--	--	--	--
	6	1	2	2	3	14	--	--
	7	1	1	1	2	3	9	17
	8	1	1	1	2	9	14	--
	9	--	--	--	--	--	--	--

STORAGE/HOLDING CONDITIONS

This discussion of storage conditions will be limited to the storage/holding temperature, and the time/temperature involved in cooling of cooked items, and the relative humidity to which the food or packaging material may be exposed. Other factors that may be included as important considerations for storage, such as the effectiveness of the packaging material at conserving certain characteristics, are discussed in other sections of this chapter.

When considering growth rate of microbial pathogens, time and temperature are integral and must be considered together. As has been stated previously n this chapter, increases in storage and/or display temperature will decrease the shelf life of refrigerated foods since the higher the temperature, the more permissive conditions are for growth. At the same time, those foods that have been cooked or re-heated and are served or held hot may require appropriate time/temperature control for safety. For example, the primary organism of concern for cooked meat and meat-containing products is C. perfringens. Illness symptoms are caused by ingestion of large numbers (greater than 108) of vegetative cells. The organism has an optimal growth range of 43 - 47 °C (109-116 °F) and a growth range of 12-50 °C (54 - 122 °F). Generation times as short as 8 min have been reported in certain foods under optimal conditions. Thus time/temperature management is essential for product safety.

The literature is replete with examples of outbreaks of foodborne illness that have resulted from cooling food too slowly, a practice that may permit growth of pathogenic bacteria. Of primary concern in this regard are the spore-forming pathogens that have relatively short lag times and the ability to grow rapidly and/or that may normally be present in large numbers. Organisms that possess such characteristics include C. perfringens, and Bacillus cereus. As with C. perfringens, foodborne illness caused by B. cereus is typically associated with consumption of food that has supported growth of the organism to relatively high numbers. The FDA "Bad Bug Book" notes that "The presence of large numbers of B. cereus (greater than 106 organisms/g) in a food is indicative of active growth and proliferation of the organism and is consistent with a potential hazard to health". In this case, the time and temperature (cooling rate) of certain foods must be addressed to assure rapid cooling for safety.

The effect of the relative humidity of the storage environment on the safety of foods is somewhat more nebulous. The effect may or may not alter the aw of the food. Such changes are product dependent. The earlier discussion on aw and its effect on microorganisms in foods provides some background information. In addition, the possibility of surface evaporation or condensation of moisture on a surface should be considered.

Generally, foods that depend on a certain aw for safety or shelf life considerations will need to be stored such that the environment does not markedly change this characteristic. Foods will eventually come to moisture equilibrium with their surroundings. Thus, processors and distributors need to provide for appropriate storage conditions to account for this fact.

Packaging, as discussed previously in this chapter, will play a major role in the vulnerability of the food to the influence of relative humidity. But even within a sealed container, moisture migration and the phenomenon of environmental temperature fluctuation may play a role. It has been observed that certain foods with low aw can be subject to moisture condensing on the surface due to wide environmental temperature shifts. This surface water will result in microenvironments favorable to growth of spoilage, and possibly pathogenic, microorganisms. As a general guideline, the product should be held such that environmental moisture, including that within the package, does not have an opportunity to alter the aw of the product in an unfavorable way.

PROCESSING STEPS

The current definition of "potentially hazardous foods" considers the effect of processing in much the same way that it considers pH and aw: it divides foods into two categories. Low-acid canned foods in a hermetically-sealed container do not require temperature control for safety. This rigid definition fails to address less processed foods, in less robust packaging, which still would not require temperature control for safety. Consider a baked product, such as a pie, with a pH of 5.5 and aw of 0.96. Since this product is baked to an internal temperature >180 °F (82 °C) to set the product structure of the pie, it will not contain any viable vegetative pathogens. Any pathogenic spores that survive the baking process will be inhibited by the pH and aw values listed above. If the product is cooled and packaged under conditions that do not allow recontamination with vegetative pathogens, the product is safe and stable at room temperature until consumed, or until quality considerations (that is, staling) make it unpalatable.

Scientifically sound criteria for determining whether foods require time/temperature control for safety should consider 1) processes that destroy vegetative cells but not spores (when product formulation is capable of inhibiting spore germination); 2) post-process handling and packaging conditions that prevent reintroduction of vegetative pathogens onto or into the product before packaging; and 3) the use of packaging materials that while they do not provide a hermetic seal, do prevent reintroduction of vegetative pathogens into the product.

FACTORS

INTENDED END-USE OF PRODUCT

In addition to carefully assessing how the product is produced and distributed, it is important to consider how the food will ultimately be prepared, handled, and/or stored by the end user. A food product that does not require time/temperature control for safety at one point in the food production or distribution chain may require time/temperature control at another point, depending on its intended use. For example, a thermally processed food that is hot-filled into its final packaging may not require refrigeration if spore-forming pathogens are not capable of outgrowth. However, once the food item is taken out of its original packaging, it may require time/temperature control for safety if the product is likely to be recontaminated during its intended use.

PRODUCT HISTORY AND TRADITIONAL USE

The panel struggled with the concept of product history and traditional use as a means to determine the need for time/temperature control for safety. For example, there are foods which have a long history of safe storage use at ambient temperatures, yet have formulations, pH, and aw that would designate them as "temperature controlled for safety" (TCS) foods. Paramount among them is white bread, but products such as intact fruits and vegetables, other breads, bottled waters, and some processed cheeses have a history of being stored and used at ambient temperatures with no public health impact.

In addition, moisture protein ratios (MPR) for shelf-stable fermented sausages were developed to ensure process control values for these sausages that also have a traditional history of safety as a non-TCS food. Moreover, an evaluation of the food characteristics provides a scientific explanation for the products to be safely stored at ambient temperatures.

For example, baking of bread controls the growth of pathogens in the interior, and the low aw precludes the growth of pathogens on the outer surface, so that it can be stored safely at ambient temperatures. Clearly these products' traditional uses and histories provide a valid justification for a decision to be made based on history. Care must be observed, however, as this traditional history can be influenced by the intrinsic and extrinsic factors and any changes in product end- use, processes, formulation, physical structure, processing, distribution, and/or storage. Changes in any of these parameters may invalidate the sole use of history as a basis for decisions on whether a food needs temperature control for safety.

The panel recognizes that the use of history as a factor to decide whether a product needs time/temperature control for safety can be subjective. As a guidance, one should determine whether the food in question or any of its ingredients have been previously implicated as a common vehicle of foodborne disease as a result of abuse or storage at ambient temperature. Of particular importance are the microbiological agents that may be of concern based on food formulation, or that may be responsible for illnesses associated with the food and the reported contributing factors that have led to documented illnesses.

Has adequate temperature control been clearly documented as a factor that can prevent or reduce the risk of illness associated with the food? As intrinsic or extrinsic factors change (for example, MAP or greatly extended shelf life), historical evidence alone may not be appropriate in determining potential risk. Therefore, for a product to be identified as non-TCS based on history and traditional use, the intrinsic and extrinsic factors affecting microbial growth need to have remained constant. Lastly, product history alone should not be used as the sole factor in determining whether a food needs time/ temperature control for safety. This decision requires a valid scientific rationale such as that provided above for white bread.

INTERACTIONS OF FACTORS

Traditional food preservation techniques have used combinations of pH, aw, atmosphere, numerous preservatives, and other inhibitory factors. Microbiologists have often referred to this phenomenon as the "hurdle effect". For example, certain processed meat products and pickles may use the salt-to-moisture ratio (brine ratio) to control pathogens.

USDA recognizes this strategy in designating as shelf-stable semi-dry sausages with a moisture-protein ratio of less than or equal to 3.1:1 and pH less than or equal to 5.0.

In salad dressings and mayonnaise-type products, the acid-to-moisture ratio along with pH is the governing factor for pathogen control. An acid:moisture ratio > 0.70 in combination with a pH < 4.1 is often used as the pathogen-control target level for these products. Usually, these ratios are combined with other factors such as pH or added antimicrobials to effect pathogen control (Mossel and others 1995). It is the interaction of these factors that controls the ability of pathogens to proliferate in foods.

Despite this long-standing recognition of the concept of hurdle technology (the possible synergistic effect of combining different inhibitory factors), the current definition of potentially hazardous foods only considers pH and aw independently, and does not address their interaction. The panel believes that these interactions have to be taken into consideration.

Scientific advances in predictive food microbiology over the last two decades have repeatedly shown that different inhibitory factors that might not prevent pathogen growth when considered singly will prevent pathogen growth when used in concert.

It should be noted that this model was developed in broth with salt and pH combinations and that growth of bacteria in food systems will likely differ. Also, the salt used to control the aw results in additional microbial inhibitory effects that may be lacking if other compounds are used. The values are the time in hours needed for a 3 log increase in S. aureus concentration as a function of the pH and aw values shown.

It is clear from the numerical values shown that even though a food might have a pH of 5.0 and an aw of 0.92 (for example), after 72 h at room temperature, it may show a minimal increase in S. aureus concentration, and thus not constitute a significant risk to public health.

Models that address the interaction of other factors (for example, atmosphere, preservatives) have been published, but are not nearly as numerous as models using pH and aw. Individual companies have shown, however, that in-house models incorporating preservative effects can be useful tools in reducing the need for extensive challenge testing and assessing risk. However, a general model for foods to cover all interactions of atmospheric gases and/or preservative combinations with pH and aw does not currently exist.

In order to design effective combinations of factors, an understanding of the pathogen (vegetative or spore-forming) and of the mechanisms by which individual factors exert their impact are necessary.

3

Germination Testing

Of all the quality measurements of seed lots, none is more important than the potential germination of the seeds. The main aim of a laboratory germination test is to estimate the maximum number of seeds which can germinate in optimum conditions. The use of standardized ideal conditions in the laboratory such as those prescribed by ISTA ensures that results obtained from a given seed lot in one laboratory should be identical with those obtained from any other laboratory in the same or in other countries. A common standard for assessing germinative potential is very valuable for seed moving in international trade. On the other hand, it is clear that results obtained under ideal controlled conditions in the laboratory are not directly applicable in the field nursery, where only limited control over environmental conditions is possible. Each nurseryman should apply his own correction factor, derived from experience over the years, to convert the germination potential of a seed lot, as determined from laboratory tests, to the actual field germination he can expect under his own local conditions.

At the other extreme, a nurseryman may prefer to carry out his own germination test in his own nursery in advance of large scale sowing. The results of the test should be directly applicable to the later sowings of the same seed lot in the same nursery, but would not be applicable to other nurseries. However, there is sometimes insufficient time for germination tests in advance of the main sowing and managers of large operational nurseries may be reluctant to involve themselves in small scale research.

Between the two extremes lies the case of the small silvicultural research station which lacks the laboratory facilities to comply with ISTA prescriptions for testing, but which carries out germination tests in the nursery on bulk seed supplies before distributing them to afforestation projects throughout the country. Here again the local forester or nurseryman must apply his own correction factor to convert actual germination figures from the research nursery to expected germination figures in his large operational nursery.

The following account is largely based on that of Turnbull. It is in conformity with ISTA rules (1976), but some of the principles are equally applicable if lack of equipment or trained staff makes it necessary to use simpler methods. Germination is defined as the emergence and development from the seed embryo of those essential structures which are indicative of the seed's capacity to produce a normal plant under favourable conditions (Justice 1972, ISTA 1976). Germination is expressed as the percentage of pure seeds which produces normal seedlings or as the number of seeds germinating per unit weight of the sample.

In the laboratory the environmental conditions, including moisture, temperature, aeration and light, must not only be specific enough to initiate germination but also favourable for the development of the seedlings to a stage where interpretation of normal and abnormal types may be made.

With a few exceptions, all germination tests should be made with pure seeds separated by the purity test. The pure seed must be well mixed and counted at random into replicates. The seeds should then be spaced uniformly on the test substrate. Normally one test consists of 400 seeds in 4 replicates of 100 seeds each, but if 100 seeds overcrowd the test substrate, the replicates may be broken down into a larger number of smaller replicates of 50 or 25 seeds each. A general recommendation is to leave 1.5 to 5 times the normal seed width or diameter between seeds, in order to discourage the spread of fungal moulds.

The exceptions are very small-seeded species in which it is impossible (some species of Eucalyptus) or very difficult (Alnus, Betula, Populus,Salix) to separate the seeds from the accompanying inert matter or chaff. In these cases the test is made with the same number of replicates but the replicates are of equal weight rather than an equal number of seeds.

GERMINATION EQUIPMENT

The kind of germinator can be selected according to the type and amount of seeds being tested and it is acceptable providing it can give good control of prescribed temperature, moisture, and light conditions.

Germinators range in size from small or portable individual units through cabinets of various sizes and large germination tables of the Jacobsen Copenhagen type, to walk-in germination rooms. The major types recommended by ISTA are as follows:

Jacobsen and Rodewald Apparatus

In Europe, a germinator called the Jacobsen apparatus or Copenhagen tank is commonly used. It consists of a bath containing water, the temperature of which may be controlled by thermostats. The seeds are distributed on top of paper and placed on metal or glass strips suspended about 5–7 cm above the bath; paper or cotton wicks lie beneath the substrate and pass through slots into the water below. The moisture content of the substrate may be adjusted by altering the water level, and so increasing or decreasing the distance between the seed bed and the water.

High humidity is maintained around the seed either by covering the entire tank with a transparent lid or placing an inverted plastic funnel with a hole in the conical end over each germination pad. The apparatus may be exposed to natural daylight but artificial light is usually desirable.

One of the disadvantages of the standard Copenhagen tank is the absence of direct temperature control at the seed bed. In the tropics it tends to overheat. Improved models have been developed, such as the one by Overaa which uses hollow stainless-steel strips, through which the water for heating and cooling is circulated.

Other disadvantages are that the equipment requires a large amount of floor space for the number of tests accommodated. Also, the handling of covers and substrata appears to be more cumbersome than moving a lightweight tray with paper substrate.

The Rodewald apparatus consists of a zinc box covered with glass in which the seeds are exposed to direct or diffuse light. The bottom of the apparatus contains water over which a layer of moist sand is placed on a tray. The seed bed consists of unglazed porcelain dishes placed in the moist sand, or the dishes may be placed directly in the water. The temperature of the water is controlled thermostatically and the sand is moistened through wicks reaching into the water bath. A sand substrate is not desirable for species requiring alternating temperatures because of its slow adjustment to a change in temperature.

Germination Cabinet

Another common type of apparatus is the closed chamber for germinating seeds in darkness, diffused light, or direct light. This type of germinator often consists of a double walled cabinet, suitably insulated against temperature changes by an air jacket or layer of insulating material. It is equipped with appropriate tray slides to accomodate the type of germination trays preferred by individual laboratories. The modern germination chamber has provision for both heating and cooling. Usually, water is chilled and passed between the walls of the chamber or through cooling pipes placed along the inner walls of the chamber. Either the air or a reservoir of water in the bottom of the chamber may be heated. In such cabinets the temperature can be regulated at any desired setting between approximately 8°C and 40°C.

Inexpensive germination cabinets are sometimes constructed locally. Gupta and Kumar (1977) describe the low cost germinators in use in the Seed Testing Laboratory in Dehra Dun, India. The outer wall of the cabinet is constructed of teak, the size 195 × 70 × 40 cm, and the electric controls include a thermostat, hot air blower, a separate air circulator and a time switch to control illumination. Glass wool is used for insulation between the inner and outer walls. Satisfactory performance is obtained at temperatures varying from room temperature up to 45° C (± 1° C).

A germination cabinet should meet as far as possible the following requirements (Oomen and Koppe 1969):

- *Air humidity*: As high as possible but preferably never lower than 90%, to prevent the seed substrate from drying up;
- *Air temperature*: Adjustable between 10°C and 35°C, the temperature all over the working section of the cabinet over a number of days to be uniform in each of the temperature regimes, within plus or minus 1°C;
- *light*: Uniform illumination of the trays, intensity 750 to 1250 lux at the level of the seed;
- *Air movement*: As low as possible to prevent the seeds from drying up;
- *Fresh air supply*: small, of the order of one air change per hour, which should be sufficient to remove the carbon dioxide produced by the germinating seeds;
- Day and night cycling: A rapid initial temperature drop to be achieved in 30 minutes during the day to night change, reaching the ultimate night temperature in one hour, a similar requirement to hold for the night to day change;
- *Condensation*: Not allowed.

However, when cabinets operate with alternating temperatures it is very difficult to maintain the high humidity and the substrates dry out; this necessitates frequent watering, which adversely affects the germination results and adds to the manhours spent on the test.

There are many modifications of the germination cabinet and for detailed descriptions of some of them see Justice (1972) and Oomen and Koppe.

Room Germinator

When large numbers of tests are made, whole rooms may be used as germinators, with temperature, humidity and light controlled.

The room germinator is a modification of the cabinet. It is constructed on the same principle as the cabinet, but is large enough to permit workers to enter it and place the tests along either side of a central passageway. Usually, fans are required to prevent stratification of temperature and special equipment is necessary to maintain a high relative humidity.

Another modification is the combination room-cabinet germinator. The temperature of the entire room is kept as low as the lowest temperature required. Germination cabinets are placed in this room and heated, individually, by electricity to maintain the various temperatures required. Either constant or alternating temperatures may be obtained by this type of germinator.

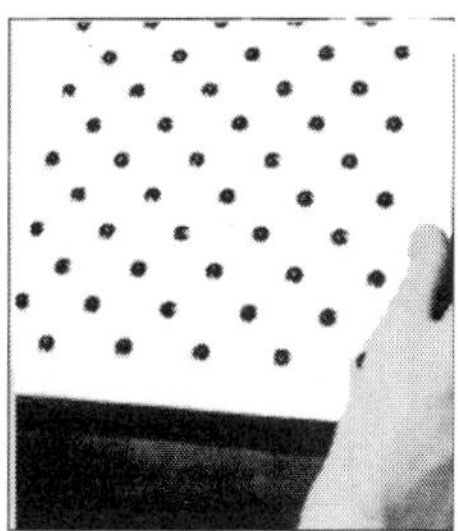

Fig. Counting board with seeds of *Celtis laevigata*. Seeds are spread on the top board to place one seed in each hole. The spring-held top board is then moved to the right until its holes line up with another set in the bottom board; the seeds then drop through.

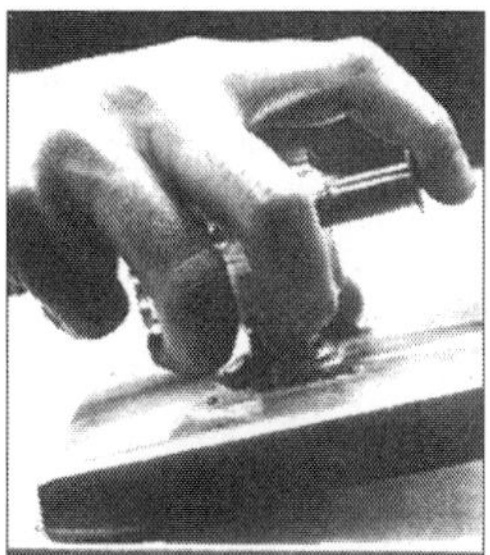

Fig. Counting head on a vacuum seed counter. A pump pulls a vacuum through the line and the hollow counting head that has 50 or 100 tiny holes on its bottom surface. When one seed per hole is attached, the vacuum can be released with the finger plunger to drop the seeds.

Fig. Seed germination equipment at the Division of Forest Research, CSIRO Canberra (A) Open seed germination cabinet (B) Group of cabinets.

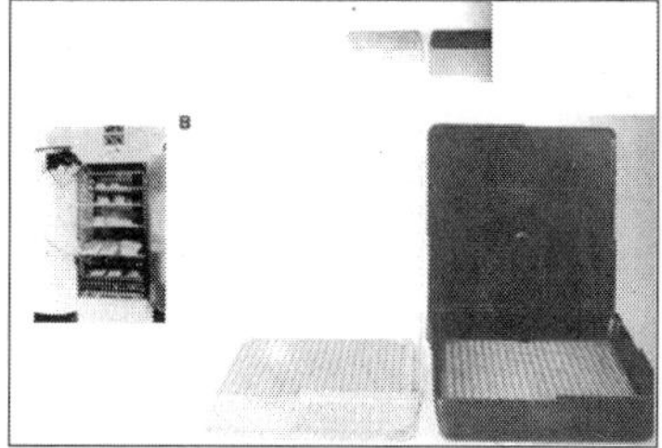

Fig. Conviron G30 germinator, with over 95 per cent relative humidity and programmable temperature and photoperiod within 24 hours, as used at Petawawa National Forestry Institute, Canada. (Canadian Forestry Service) A. Clear and black germination boxes developed for seed testing

Fig Copenhagen tank and rolled filter paper for germination tests.

Portable germination boxes. A simple and versatile germinator consists of a number of transparent plastic boxes with lids, which can be stacked one upon another.

Robbins (1984) describes the ideal container as being:

1. Rectangular and stackable, for economony of space;
2. Large enough for adequate spacing of at least one replicate of seeds (100, 50 or 25 seeds, depending on seed size);
3. Sufficiently deep to allow for the required substrate depth and to permit development of the seedling for proper assessment;
4. Provided with a wellfitting lid, to maintain a high moisture content of the substrate and surrounding air;
5. Easy to sterilize by heat or chemical treatment;
6. Transparent (at least the lid), to allow light if required for germination and subsequent development of the seedlings.

In Honduras the size of box used is 178 × 117 × 72 mm deep and each box contains one replicate of 100 pine seeds. This leaves a space of at least the width of one seed between each seed. For larger-seeded species there are fewer seeds per replicate box. Any of the common substrates described, *e.g.* filter paper, sand, may be used. By adding a suitable quantity of water to the substrate at the beginning of the test and then keeping the boxes covered with their lids at all times except during assessment and removal of seedlings, it is possible to maintain a high and constant moisture content in the substrate and the air inside the boxes, without the need for further additions of water or for control of humidity in the atmosphere outside the boxes. Where filter paper or blotting paper is used as the substrate, it can be kept permanently moist by placing it on a raised platform over a reservoir of water which is fed to the substrate by wicks. This arrangement resembles, in miniature, the Jacobsen tank. The boxes may be placed inside an incubator where temperature and light can be controlled in accordance with *e.g.* ISTA recommendations or, if no incubator is available, they can be stored in ambient room conditions of temperature and light. In either case the closed boxes should provide optimum humidity conditions for the germinating seeds.

Researchers at the Petawawa National Forestry Institute in Canada have developed a germination box, moulded out of light, unbreakable and heat-resistant polycarbonate plastic, which is large enough to contain 4 replicates of

100 seeds each of pines or similarly sized seeds. The box is 28 cm long × 24 cm wide. The base is 5 cm deep and the lid 1 cm deep, but the special universal locking mechanism makes it possible to fit two bases together; alternative combined depths of 6 and 10 cm are thus available for use according to the characteristics of the species under test. A perforated false bottom on eight legs 1 cm long supports the germination substrate and the space below can be used as a water reservoir if desired. The material is available in either clear or black plastic for germination in light or dark conditions; to ensure complete darkness it is necessary to seal the junction between base and lid with opaque tape. Four aeration holes on the side wall of the bottom component are optional in the clear plastic type.

Tests of this germination box gave results closely comparable with those from the much smaller, and thus less convenient, petri dishes. It was found that loss of moisture over a four week period was least in boxes which had the water reservoir filled at the start of the test and which had no aeration holes. Both clear and black were equally good in this respect, but growth of mould was more severe in the (sealed) black than in the clear model. A substrate of blotting paper over Kimpak (cellulose paper) lost moisture more slowly than Kimpak alone.

Selection of a germinator. Most forest seed laboratories with a moderate through-put of samples use either the cabinet-type germinators or germination tables. A report by Boeke et al. (1969) recommends the exclusive use of cabinettype germinators for small seed testing laboratories. The advantages of using cabinets instead of germination tables of the Copenhagen type are a saving in space and, in well constructed cabinets, the more precise control of temperature and humidity (Oomen and Koppe 1969). It should be noted that a single-level Copenhagen tank requires five to eight times the floor space occupied by a cabinet of equal capacity (Boeke et al. 1969). Where research is being carried out on germination conditions, or where testing covers a wide variety of seeds, the flexibility provided by several cabinets operating at different temperatures and/or light conditions can be very advantageous.

For nurseries or small research institutes concerned with carrying out simple tests of germination without maintaining strict control of temperature and light, the flexibility, storability and cheapness of plastic germination boxes have much to recommend them. They can also be used in combination with incubators in which light and temperature are controlled.

GERMINATION CONDITIONS

The optimum conditions for different stages of germination and seedling growth are not identical and may even vary for different seeds within the same seed lot. The aim of much seed testing research has therefore been to determine a combination of conditions which will give the most regular, rapid and complete germination for the majority of the same species.

Substrate

Soil is rarely used as a substrate for germination tests because each sample will vary greatly in physical, chemical and biological properties. While the germination results might be more comparable with field conditions, the lack of reproducibility and difficulty in comparing tests of different seed lots render it unsuitable. Artificial media are much more easily standardized.

Most laboratory tests on small seeded species are made on paper. Other materials used include sand, granulated peat moss and expanded mica (Vermiculite and Terralite).

The main requirements for the substrate are:

- Non-toxic to the germinating seedlings
- Free of fungi and other micro-organisms
- Of porous texture to enable adequate aeration and moisture for the germinating seeds.

The choice of media on which the seeds are placed for germination depends on the equipment, the species, the working conditions and the experience of the operator. It is prescribed in the ISTA Rules for a number of forest trees.

Filter paper, heavy box paper, or any other absorbent paper may be linked with a filter paper or cotton wick dipped into water, or placed on top of a layer of sand or vermiculite. Cellulose paper is becoming more commonly used as a germination medium because it is easier to handle than sand, but still permits penetration of the radicle, thus providing for better evaluation of abnormal germination. Also moisture content does not tend to become stratified within the paper medium as it does with sand.

The best paper substrates are germination blotters, paper towelling, laboratory filter paper and creped cellulose paper wadding. Any paper substrates should be checked for presence of toxic chemicals. The 1976 ISTA rules give detailed specifications for paper and towels, including weight, bursting strength, capillary rise and acidity.

Seeds without a specific light requirement may be placed on top of paper or between folded paper. Folded paper increases the surface contact area between the seeds and the moisture supply. Large seeds may be rolled in paper towels which are then placed in an upright position; this allows the roots to grow downwards and avoids tangling of the roots. Rolled or folded paper packs may be kept moist without the necessity of a wick dipping into water. Rolled seeds may be stored on glass plates above, but not in contact with, a water bath and covered by a polythene sheet. Any rolls which show signs of drying out may be sprinkled with water. Use of rolled paper is a very rapid and convenient method but may cause curling of the radicles. This is of no concern if the germinated seeds are to be discarded after testing. For operational germination of large quantities of seed, however, the slightly slower folded paper method is preferable. This is used with success as standard practice for Pinus and Eucalyptus spp. in Thailand. Sand is not suitable for very small seeds because they are difficult to locate, but it is widely used for larger seeds. Sand can be sterilized and fungi develop on it less freely than on paper. It also provides a good contact between the seed and moisture because the seeds can be pressed into the medium. A commonsense rule is to cover the seed with a thickness of sand not less than the length of the seed on its longest axis. ISTA recommends a thickness of 1 – 2 cm according to seed size and prescribes that sand particle sizes should be between 0.05 and 0.8 mm and pH between 6.0 and 7.5.

Sand as a germination substrate is preferred for tree species that have a longer germinative period, for instance, Rosa spp., Pinus caribaea, P. elliottii, and P. palustris, or with larger seeds, for example, P. pinea and Quercus spp. A sand-perlite medium permits leaching of the water which can be important when testing repellent-coated seed, but may be a disadvantage when testing seeds which are sensitive to drying.

Moisture and Aeration

The moisture level of the substrate has been suggested as one of the major causes of variation in seed research results. The ISTA Rules specify that a sand substrate be moistened depending on its characteristics and the size of the seed to be tested and suggest that a moisture level of 50 – 60 per cent of the water-holding capacity of the sand is suitable for several groups of agricultural seeds. A paper substrate must not be so wet that a film of water forms around the seed. Belcher, working with Pinus spp., concluded that most species have considerable tolerance to diverse moisture levels and the major variations in germination were due to dry conditions. He found some species to be sensitive to dry conditions and others to very wet conditions.

As a generalization, the substrate must be moist enough at all times to supply the necessary moisture, but excessive moisture will restrict aeration. All tests must be examined daily to ensure that the moisture content of the substrate is near optimum. The water should be reasonably free of impurities.

Temperature Control

In laboratory seed testing the important temperature is that at the level of the seed. The temperature will vary for different species and appropriate values are listed for a large number of tree species in the International Rules for Seed Testing. In addition to temperature, it indicates prescribed substrates, light, counting periods and pretreatment.

Temperature is one of the most critical factors in the laboratory germination of seeds and must be regularly checked. When alternating temperatures are required the test is usually held at the low temperature for 16 hours and at the high temperature for 8 hours each day. Temperature of 20° and 30°C alternating are commonly prescribed for tree species. Although natural fluctuations between day and night temperatures are less in lowland moist tropical forests than in other forest types, alternating temperatures may still affect germination of tropical species. In one experiment of Terminalia ivorensis in Nigeria, alternating temperatures of 34° C and 24° C gave 93 per cent germination in 41 days compared with 27 per cent at a constant temperature of 30° C, both with continuous light.

Light

Light is required for germination in many tree seeds. Fluorescent light is as effective as natural daylight and is preferred in testing, because the wavelength and intensity can be standardized, within limits. Cool white fluorescent lamps are recommended because of their light quality and low heat emission: Light should be evenly distributed over the tests, with an intensity within the range 750 – 1250 lux. Seeds should be subjected to light for only part of the test period, 8 hours in every 24 hours is usual but a longer or shorter period may be beneficial to seeds of some species.

Cotrolling Fungi in Germination Tests

Laboratory practices that will minimize the spread of fungi include proper spacing of seeds, control of temperature, removing decayed seeds, proper aeration, and keeping the substrate only just wet enough to permit germination. Sterilization of laboratory equipment and the periodic disinfection of germination cabinets and other apparatus will assist.

As a general rule, seed is not disinfected because seeds which decay are usually of poor quality. Magini (1962) states that a number of experiments carried out with different methods of disinfecting the seed just before or during germination have shown no reliable improvement in germination percentage. A small quantity of fungicide added to the water used to moisten paper rolls has, however, been found beneficial in testing conifer seeds in Denmark. In Australia spraying of Karathane fungicide at a concentration of 0.8 g in one litre of distilled water has also proved effective when testing eucalypt seed on moist filter paper over vermiculite.

GERMINATION CONDITIONS FOR SELECTED SPECIES

Conditions prescribed by CSIRO and ISTA for germination testing of selected tropical, sub-tropical and temperate species are tabulated in Tables 9.1, page 213 and 9.2, page 215 respectively.

Notes for table:

1. Temperature recommendations separated by a semi-colon indicate that the (constant) temperatures have been found satisfactory. Alternating temperatures are not necessary.
2. Unless otherwise specified, normal substrate used for testing

eucalypts by Division of Forest Research, CSIRO, Canberra is moist filter paper over vermiculite in a petri dish. Where "vermiculite only" is recommended, the filter paper should be omitted.

3. A temperature shown in parenthesis *e.g.* (25) indicates that satisfactory results have been obtained with that temperature but a full range of temperatures has not been tested.

Species	Mean No. of Viable Seeds per g of Seed and Chaff	Seed Testing Recommendations				Special Recommendations 2)
		Weight of Replication (g)	Temperature (°C) 1)	First Count (days)	Final Count (days)	
E. brassiana	340	0.15	25	7	14	
E. camaldulensis	670	0.10	30	5	10	Light essential
E. citriodora	110	0.50	25;30	5	14	Vermiculite only
E. cloeziana	130	0.40	25	7	28	Vermiculite only
E. deglupta	4,000	0.01	35	5	14	Vermiculite only
E. globulus subsp. globulus	75	0.70	25	5	14	
E. grandis	650	0.10	25	5	14	
E. microtheca	380	0.15	35	3	14	Light essential,
E. regnans	180	0.30	15	10	21	Vermiculite only or
E. saligna	540	0.10	25	5	14	stratify 3 weeks, then
E. tereticornis	600	0.10	25;30;35	5	14	20°C
E. urophylla	460	0.10	(25) 3)	5	14	

This table indicates permissible substrates, temperatures, duration and additional directions, including recommended special treatments for dormant samples. For species in section 1, the methods in columns 2 – 6 are prescriptive and no others may be used. Where two figures are shown under temperature, *e.g.* 20 – 30, this indicates daily alternation of temperatures. The lower temperature should be maintained for 16 hours and the higher for 8 hours, in each 24 hour period.

For species in section 2, other methods may be used provided that the method is indicated on the International Analysis Certificate. For certain species indicated in column 7, duplicate tests (with and without prechilling) are necessary; the germination phase of these should run concurrently.

The less desirable methods are placed in brackets in the Table.

The abbreviations have the following meanings:

- TP - top of paper
- BP - between paper (including rolled towels and pleated paper)
- S - sand
- TS - top of sand
- L - light essential
- TT - topographical tetrazolium test

	Species	Prescriptions for:					
		Substrata	Temperature (°C)	Light	First Count (days)	Final Count (days)	Additional directions including recommendations for breaking dormancy
	1	2	3	4	5	6	7
	Secrtion 1 (Prescriptive)						
	Acacia spp.	TP	20–30(20)	L	7	21	(1) Pierce seed, chio or file off fragment of testa at cotyledon end and soak 3 hrs. or (2) (Soak seed 1 hr. in concentrated H_2SO_4, wash seed thoroughly in running water after acid treatment).
	Ailanthus altissima	TP	20–30	-	7	21	Removal of pericarp after soaking for 24 hrs. may speed up germination.
1)	Alnus spp.	TP	20–30	L	7	21	(May be tested also by using 4 weighed replicates of 0.10 to 0.25 gm each, according to species)
	Cedrela spp.	TP	20–30	L	7	28	
	Cryptomeria japonica	TP	20–30	L	7	28	
	Cupressus sempervirens	TP	20	L	7	28	
1)	Eucalyptus camaloulensis	TP	30	L	3	14	Use 4 weighed replicates of 0.10 gm each
	Liquidambar styraciflua	TP	20–30	L	7	21	Sensitive to drying in test
	Nothofagus obliqus	TP	20–30	L	7	28	No prechill and prechill 28 days at 3–5 ° C. Double tests.
	Pinus caribaea	TP	20–30	L	7	21	
	P. elliottii	TP	22;20–30	L	7	28	
	P. pinaster	TP	20	L	7	35	(1) No prechill and prechill 28 days at 3–5 ° C. Light for no more than 16 hrs. per day. Double tests. (2) (Use TT)
	P. radiata	TP	20	L	7	28	
	P. taeda	TP	22;20–30	L	7	28	
	Quercus spp.	TS (S)	20	-	7	28	Soak seed for up to 48 hrs. and cut off 1/3 at scar end of seed and remove testa.
	Robinia pseudoacacia	TP	20–30	L	7	14	(1) Pierce seed or chip or file off fragment of testa at

- 1) - reference to weighed replicates in Eucalyptus and other genera has been deleted by a 1981 ISTA amendment. A practicable and up-to-date guideline to testing eucalypts is contained in appendix 3 of Boland et al. 1980. Data from that appendix for a few of the more important species is reproduced in Table.

Evaluation

A seed is considered to have germinated after the emergence and development from the seed embryo of those essential structures which are indicative of the seed's capacity to produce a normal seedling under favourable conditions. Abnormal seedlings are not included in the germination count because they rarely survive to produce plants.

The ISTA Rules (ISTA 1976) recognize four groups of abnormal seedlings:

a. Damaged seedlings,
b. Deformed seedlings,
c. Decayed seedlings,
d. Seedlings with unusual hypocotyl development.

These groups and their characteristics are defined in detail in the ISTA Rules. In a laboratory test the majority of the normal seedlings are usually removed at the interim counts, but the assessment of many of the doubtful and abnormal seedlings must be left until the end of the test, to ensure that slower growing but otherwise normal seedlings are not incorrectly classified.

For many species the initial count is carried out one week after starting the test and assessments may be carried out at weekly intervals until the test is ended. More frequent assessment is needed if a more precise picture is needed of the speed of germination. At the end of the test period, all remaining ungerminated seeds should be cut and examined, and the number of fresh, firm and possibly viable seeds recorded (Bonner 1974). At the same time observations may be made on the condition of unsound ungerminated seeds *e.g.* an abnormally high incidence of insect - damaged or mechanically damaged seeds which would indicate the need for improvements in seed hygiene or seed processing methods. The record of the germination test commonly shows the percentage of germinated and the percentage of ungerminated but apparently sound seeds separately, *e.g.*

Germination %	=	82 %
Sound ungerminated seeds	=	6 %
Viability %	=	82 % + 6 % (= 88 %)

It can be seen from Tables that the duration of a test commonly lasts from two to five weeks, but this does not allow any period for pretreatment to overcome dormancy. Many species require no pretreatment and some types of seedcoat dormancy can be successfully treated in a matter of hours. But the prescribed pretreatment of Tectona takes 18 days and some prescribed prechilling treatments 3 – 9 months. When planning a testing programme it is necessary to take account of pretreatment as well as testing time. In extreme cases it may be necessary to replace the germination test by one of the indirect tests of viability described below.

Fig. Acorns of *Quercus Alba* Germinating on Kimpak in USA. Note spacing between seeds.)

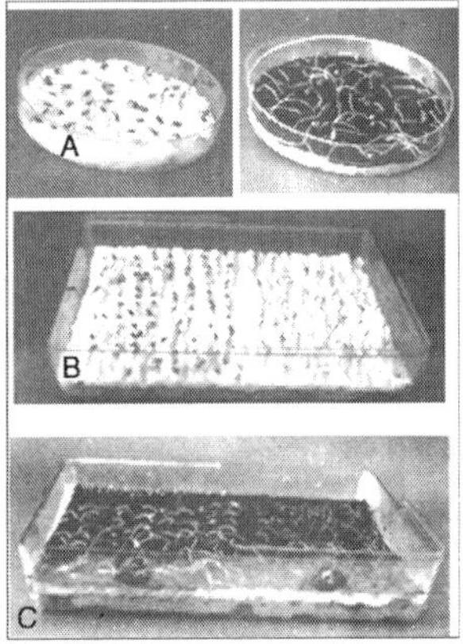

Fig. Germination of Douglas-fir and Lodgepole Pine seeds on Kimpak (A left) and blotter/Kimpak (A right) in Petridishes (A) and Canadian clear Germination Boxes (B-C).

GERMINATION ENERGY

Germination Energy has been defined in more than one way: (1) The per cent, by number, of seeds in a given sample which germinate within a given period (defined as the energy period) *e.g.* in 7 or 14 days, under optimum or stated conditions or (2) The per cent, by number, of seeds in a given sample which germinate up to the time of peak germination, generally taken as the highest number of germinations in a 24 hour period.

In both the above definitions the length of the energy period is considerably shorter than the full test period prescribed by ISTA. Germinative energy is a measure of the speed of germination and hence, it is assumed, of the vigour of the seed and of the seedling which it produces. The interest in germinative energy is based on a theory that only those seeds which germinate rapidly and vigorously under the favourable conditions of the laboratory are likely to be capable of producing vigorous seedlings in field conditions, where weak or delayed germination is often fatal. Little experimental evidence has been published on this theory, but excessively delayed germinants must get eliminated automatically from the nursery, either because they are suppressed by older and more vigorous competitors or because, if transplanting has already been completed, they are not worth the trouble of a special supplementary transplanting. An example of calculating germination energy is given on pp. 234–236.

Another method of comparing germination energy of different seed lots is to record "Rate of Germination" *i.e.* the number of days required to attain 50 per cent of germination capacity. The shorter the period, the greater the germination energy. Yet another method is to evaluate the stage of development of germinated seeds and classify them into a number of development or vigour classes. For example Wang used seven classes for germinated normal seeds of Picea glauca, in addition to ungerminated and abnormally germinated seeds; they ranged from seedlings with healthy root, fully developed hypocotyl and seedcoat completely shed to seeds with seedcoat burst open but as yet no radicle emergence.

GERMINATION VALUE

The concept of Germination Value, as defined by Czabator, aims to combine in a single figure an expression of total germination at the end of the test period with an expression of germination energy or speed of germination. Total germination is expressed as (final) Mean Daily Germination (MDG), calculated as the cumulative percentage of full seed germination at the end of the test, divided by the number of days from sowing to the end of the test. Speed of germination is expressed as Peak Value, which is the maximum mean daily germination (cumulative percentage of full seed germination divided by number of days elapsed since sowing date) reached at any time during the period of the test.Germination Value (GV) can then be calculated from the formula.

$$GV = \text{(final) } MDG \times PV$$

Germination Value, as an integrated measure of seed quality, has been used by several tropical seed workers *e.g.* for Terminalia ivorensis and for Pinus kesiya.

An alternative method of calculating Germination Value has been proposed by Djavanshir and Pourbeik, who found that it was more closely related to survival of plants in field nurseries than was Czabator's method, in the case of Pinus ponderosa and P. eldarica in Iran. The formula they proposed was:

$$GV = (\Sigma DGS / N) \times \frac{GP}{10}$$

where:

GV	=	Germination value
GP	=	Germination per cent at the end of the test
DGS	=	Daily germination speed, obtained by dividing the cumulative germination per cent by the number of days since sowing
DGS	=	The total obtained by adding every DGS figure obtained from the daily counts
N	=	The number of daily counts, starting from the date of first germination.

The calculations required are somewhat lengthier than those of the Czabator method and, for many patterns of germination, the simpler method is likely to give sufficiently accurate comparisons between seed lots.

GERMINATION TESTING IN THE NURSERY

A good example of a suitable method for testing germination in the nursery is the procedure recommended for tropical pines in West Malaysia:

"Take 400 pure seed and divide into 4 samples each of 100 seed. Take 4 wooden or plastic boxes (opaque flexible plastic sandwich boxes are very good) of dimensions 30 × 30 cm and of depth about 10 cm and fill them slightly overfull with sieved sand to the top of the sides. Wet the sand by applying 0.5 litres of water to each box. Level the sand across the top of each box and make drills at 2.5 cm intervals and of depth 6 mm (for P. oocarpa the drills should only be 3 mm deep). Sow the seed (100) at intervals of 2.5 cm and gently press them on to the sand in the drill. Cover to a depth of 6 mm (3 mm for P. oocarpa) with sand which has been passed through a Mesh No. 12 (12/64" or 4.76 mm) but retained on a Mesh No. 8 (Size 8/64" or 3.18 mm). Lightly water with a fine mist spray. Cover with transparent polythene sheeting of 0.2 mm thickness mounted on the upper side of a square tight-fitting wooden frame of 5 cm thickness. Within 24 hours condensation of moisture will appear on the inside of the polythene cover. If moisture disappears at any time during the ensuing 7 days then give a light mist spray and close lid once more. Germination will usually begin by the 7th day and the cover can be removed. Keep sand moist. A seed is recorded as having germinated when it has reached a height of 1 cm complete with seed coat enclosing the cotyledons. Record all germination from 7th to 28th day separately for each of the four boxes. Remove seedlings as soon as they are recorded plus any diseased ones which may not yet have reached 1 cm height as they may infect others.

On the 28th day after sowing sieve the top layer of sand in each box through a No. 12 mesh screen (mesh size 12/64" or 4.76 mm) and record the number of full seeds which have not yet germinated. (Use cut test.)"

TESTING HOMOGENEITY OF GERMINATION RESULTS

The use of 4 replicates in a germination test enables the tester to measure the degree of variation in the sample. A simple method is to calculate the range of difference in germination per cent between the highest and the lowest of the sub-samples. The range can then be compared with the table published by ISTA, reproduced here as Table (it is applicable to replicates of equal numbers of seeds, not of equal weights). Provided that the actual range is less than the maximum in the table, the sample can be regarded as homogeneous and the average of the four replicates is accepted. If the actual range exceeds the maximum in the table, then a new sample must be drawn and tested. Common causes of such situations are non-homogeneity of seed and, in nursery tests, fungal or insect damage in one or more of the replicates. Retesting is also necessary when an extremely high percentage of full, ungerminated seeds are left at the end of the test (Bonner 1974). In this case retesting should be done with a different pretreatment to try to overcome dormancy and improve total germination.

Maximum Tolerated Ranges Between Replicates This table indicates the maximum range (*i.e.* difference between highest and lowest) in germination percentage tolerable between replicates, allowing for random sampling variation only at 0.025 probability. To find the maximum tolerated range in any case calculate the average percentage, to the nearest whole number, of the four replicates: if necessary, form 100-seed replicates by combining the sub-replicates of 50 or 25 seeds which were closest together in the germinator. Locate the average in column 1 or 2 of the table and read off the maximum tolerated range opposite in column 3.

Average percentage germination		**Maximum range**	**Average percentage germination**		**Maximum range**
1	2	3	1	2	3
99	2	5	87 to 88	13 to 14	13
98	3	6	84 to 86	15 to 17	14
97	4	7	81 to 83	18 to 20	15
96	5	8	78 to 80	21 to 23	16
95	6	9	73 to 77	24 to 28	17
93 to 94	7 to 8	10	67 to 72	29 to 34	18
91 to 92	9 to 10	11	56 to 66	35 to 45	19
89 to 90	11 to 12	12	51 to 55	46 to 50	20

COMBINING PURITY AND GERMINATION TESTS

Purity tests of most commercial seed lots of Eucalyptus are not made because it is difficult or impossible to separate the seed from the chaff of some species. Species in which seed and chaff are extremely similar in size, weight and colour include E. cloeziana, E. regnans and E. delegatensis. other small-seeded genera in which separation of pure seeds is difficult are Alnus, Betula, Populus and Salix. Even when separation is possible in these species, it is very time-consuming. In the seed laboratory in Canberra it takes only 6–7 minutes to set up a test of eucalypt seed with 4 weighed replicates, but to carry out a purity separation and set up a 4 × 100 seed test takes from 20 to 50 minutes, depending on the degree of difficulty in separating out the chaff (Turnbull 1983). The small size of seed also precludes a cutting test to determine the number of full but ungerminated seeds at the end of a test. For these reasons tests are best made on replicates by weight and results recorded as numbers of germinated seeds per unit weight of the impure mixture of seed and inert material. A squash test may also be used to give a rough estimate of viability.

Where seed is sown broadcast, the practising forester is primarily interested in the number of plants he can expect to get from a given weight of the seed lot he receives. Provided he is told that 1 kg of "impure" seed should produce 36,000 germinated seedlings, he is not worried whether this is caused by a combination of 90 per cent purity × 80 per cent pure seed germination or 80 per cent purity × 90 per cent pure seed germination. For local tests, therefore, if seed laboratory staff is insuffient, there is no objection to omitting the purity test even on species for which it would be practicable. In the case of direct sowing into individual containers, however, there are obvious advantages in having separate figures for purity and germination, since sowing is by numbers of (one to several) seeds per container rather than weight of seed per m^2 of bed or tray.

INDIRECT TESTS OF VIABILITY

Estimating the germination potential of a seed lot by actually germinating a sample of it is often the method most relevant to practical forestry. But the tests take several weeks to complete and for some species pretreatment may take some additional weeks or months. For this reason much research has been conducted to find other methods by which seed viability can be estimated accurately but much more rapidly than by germination testing. The following account of these methods closely follows that of Turnbull.

The objects of quick viability tests are:

- to determine quickly the viability of seeds of species which normally germinate slowly or show dormancy under the normal germination methods;
- to determine the viability of samples which at the end of the germination test reveal a high percentage of fresh ungerminated or hard seeds.

Only two methods, the topographical tetrazolium test and the embryo excision test were previously accepted by the International Seed Testing Association as official methods for some species of seeds. ISTA has recently accepted the X-ray method as a valid alternative to the cutting test for the detection of empty and insect-damaged seeds. The following tests can be applied depending on the circumstances:

CUTTING TEST

The simplest viability testing method is direct eye inspection of seeds which have been cut open with a knife or scalpel. If the endosperm is of normal colour with a well developed embryo, the seed has a good chance of germinating. This test is not very reliable. Seeds with milky, unfirm, mouldy, decayed, shrivelled or rancid-smelling embryos and abortive seeds that have no embryo can be judged as non-viable without much difficulty (Bonner 1974). But it is not possible to distinguish moribund, recently dead or recently injured seeds which still appear the same as sound ones. The cutting test, as already mentioned, is used at the end of a germination test to determine the apparent viability of ungerminated seeds; it is also a useful tool in estimating the size and maturity of the seed crop before collection and the efficiency of methods used in processing.

In the philippines good correlation has been found between cutting and germination tests in fairly large-seeded species such as Leucaena,Intsia bijuga and Lagerstroemia speciosa, but germination per cent was consistently 10–20 per cent less than the percentage of sound seeds on cutting test.

TOPOGRAPHICAL TETRAZOLIUM TEST

The tetrazolium method is only one of a number of biochemical tests which have been developed for seed testing. The various tests have been briefly reviewed by Moore. The tetrazolium test was introduced in 1942 by G. Lakon in Germany.

In this method living cells are stained red by the reduction of a colourless tetrazolium salt to form a red formazan. The method emphasizes the need for a knowledge of the soundness of individual embryo parts for predicting the development of embryos into countable seedlings.

The testing procedure is described in detail in the ISTA Rules which approve the test for some species of hardwoods and conifers which germinate slowly by regular germination methods. Normal practice is to soak the seeds in water for about 20 hours, then cut or puncture the seed coat to facilitate entry of the 1 per cent aqueous solution of tetrazolium (TZ) and immerse the seeds in the dark for 48 hours. The process can be greatly speeded up by cutting through the seed at a distance of one third from the micropyle and placing it in a Vitascope vacuum machine for only half an hour. This method gives satisfactory results in Denmark but interpretation of results requires more experience in the operator than the method of seed immersion followed by excision of the stained embryo. The test is done on 4 replicates of 100 seeds each.

Justice states that, while the tetrazolium procedure is good in principle, its practical use in routine testing is limited by many problems, including: difficulty in staining of some seeds; necessity of cutting of dissecting seeds to permit observation of stained parts; poor agreement with results of germination tests in some cases, especially for seed of low germination capacity; lack of uniform interpretation of staining and difficulty in interpreting the significance of different degrees of staining; and an increase in man-hours required to test 400 seeds compared to regular germination tests.

The necessity for an experienced analyst for the successful use of this "common sense test" is admitted by Moore. There is little doubt that the test can be useful for testing the viability of certain species, providing trained staff are available to prepare the seeds and evaluate the results.

EXCISED EMBRYO TEST

By this method, the seeds are soaked for 1 – 4 days and the embryos are then excised from the seeds and placed on moist filter paper or blotter discs in petri dishes. The tests are placed in the light at a constant temperature of 20°C. The condition of the embryos is examined daily. Depending upon the species and lot differences, the tests can be terminated after only a few days, up to a maximum of 14 days, or as soon as distinct differentiation into viable and non-viable embryos can be made.

The excised embryo test is similar to germination tests in that it measures the quality of the seed by their actual germination. In addition it allows some measure of the embryo dormancy to be made, by counting those seeds which, although not growing normally, have grown slightly, remained firm and have kept their colour for the test period. The test is not valid for previously germinated seeds and must not be applied to samples which contain any dry germinated seeds. The success of the test requires considerable skill and experience in the operator and the ISTA rules restrict it to only a few species.

In a comprehensive study, Schubert compared the excised embryo method with the tetrazolium method for determining the viability of dormant tree seeds. He concluded that the tetrazolium method should receive preference over the embryo excision method but that improvements in the tetrazolium test should be made by providing for the use of bactericides and stronger reducing solutions to resolve doubts in weakly stained tissues.

RADIOGRAPHIC METHODS

Radiography was first used to determine seed quality over 70 years ago. The studies of Simak and Gustafsson highlighted the X-ray technique as a diagnostic method of tree seed analysis. The X-ray contrast method which uses various contrast or radiopaque agents was developed and applied successfully to species of Pinus and Picea.

The X-ray method permits the detection of empty seeds, mechanical damage and abnormally developed internal seed structures, measurement of the thickness of the seedcoat and assessment of the seed viability when combined with a contrast agent. The X-ray contrast method is based on the principle of semipermeability. When seeds are treated with a contrast agent, for example aqueous $BaCl_2$ or vaporous $CHC1_3$, their living tissues are able to prevent its entry due to their semi-permeability, but the dead tissues become impregnated. The impregnated tissues absorb X-radiation more intensively than the unimpregnated ones and thus appear lighter on the film than the unimpregnated ones. The contrast permits living and dead tissue to be located in the seed and an estimation of its viability. There are now possibilities of using non-toxic water, instead of toxic $BaCl_2$ or $CHC1_3$, as a contrast agent for testing seed viability.

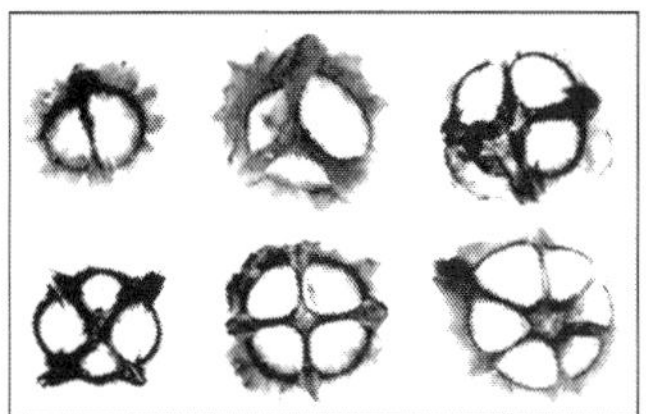

Fig. X-ray Radiograph of teak Fruits showing the Variation in the number of Locules (two to six).

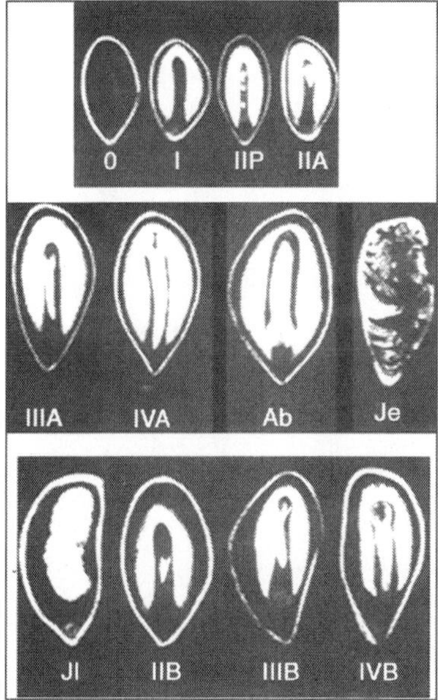

Fig. X-ray Radiographs Showing Embryo and Endosperm classes in Coniferous seeds.

O	Neither embryo nor endosperm (= empty seed)
I	Endosperm, embryo cavity developed but no embryo observed.
IIP	Endosperm, one or more small embryos the length of which does not exceed their breadth ("point embryos").
II	Endosperm, and one or several embryos, none of which is longer than half of the embryo cavity.

III	Endosperm, and one or more embryos, the longest of which measures between half and three quarters of the embryo cavity.
IV	Endosperm, with one fully developed embryo, completely or almost completely occupying the embryo cavity. Diminutive embryos rarely occur.
A	The endosperm almost fills the seed coat to capacity and easily absorbs the x-rays.
B	The endosperm fills the seed coat incompletely and is often shrunken or otherwise deformed. The x-ray absorption is inferior to that of class A.
Ab	Seed with abnormally developed endosperm or embryo.
J	Seeds damaged by insects, containing larvae (Jl) or their excrement (Je).

The development of soft X-ray equipment has greatly simplified the operation. Complicated photographic equipment is not necessary and pictures can be made with polaroid film which provides clear and detailed radiographs within 30 seconds. X-ray radiography has been successfully applied in determining the number of seeds in fruits of teak (Tectona grandis) and for studying their degrees of development. The technique has been tried on the fruits or seeds of sixty tropical forestry species and the results show that it can be reliably applied in processing such seeds.

A technique of stereoradiography as a supplement to the X-ray contrast method for use in seed quality testing has been developed by Kamra, Meyer and Wegelius. The chief advantage of stereoradiography is that it is possible for the observer to have a three-dimensional view of the object from a pair of radiographs. In this way, the exact topographical location of the contrast agent in the seed can be reliably determined. This increases the information which can be obtained from radiographs and adds to the analytical accuracy.

The X-ray method is a useful one and is likely to play an increasing role in seed testing. Earlier models of X-ray machines were expensive, but recent models, particularly from Japan are much cheaper and now cost less than a cabinet germinator. Improvements in photographic films and paper have speeded up the process and simplified the interpretation, so that technicians can be trained easily to produce consistent results. ISTA has accepted the method as a valid alternative to the cutting test for the detection of empty and insectdamaged seeds. It also shows considerable promise for distinguishing between viable and non-viable seeds among "full seeds".

Table. Germinability of fresh Collected, Undamaged Seeds (in per cent) Belonging to Different DCs, Selected by Radiography

Species	DC							
	I A	II A	III A	IV A	II P	II B	III B	IV B
Pinus sylvestris	0	50	88	99	0	5	43	68
Picea abies	0	36	82	97	0	15	71	92

For certain temperate conifers it has been possible to obtain good correlation between the Development Class (DC) of seeds, based on the development of both embryo and endosperm, and their germinability. Figure illustrates the DCs which have been defined for conifers, and Table illustrates the germinability corresponding to each class, for Pinus sylvestris and Picea abies.

HYDROGEN PEROXIDE

Hydrogen peroxide (H_2O_2) has a stimulating effect on seed germination and has been used in a rapid test for germination of several conifers in the western USA. Seeds are soaked overnight in 1 per cent H_2O_2. The seedcoat is then cut open to expose the radicle tip and the seeds put back into 1 per cent H_2O_2 in the dark at alternating temperatures (20° and 30°C). Counting and refreshment of H_2O_2 is done after 3 or 4 days and final assessment after 7 or 8 days. Radicle growth of 5 mm or more is scored "evident", 0 – 5 mm "slight" and no growth means a non-viable or empty seed. The test is quicker but less reliable than a normal germination test (usually producing a more rapid and higher final germination), slower but simpler to perform than excised embryo and easier to interpret than TZ.

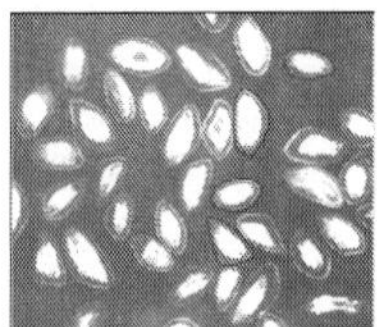

Fig. X-ray Radiograph of *Pinus Caribaea* Seed. Most seeds have very Poorly Developed Gametophyte and Embryo and are dead. A few Germinable are Indicated by the black Outlines.

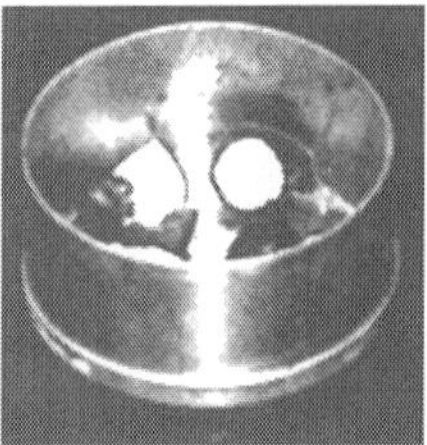

Fig. *Quercus* Seeds cut in halves for oven Drying in Moisture Determination.

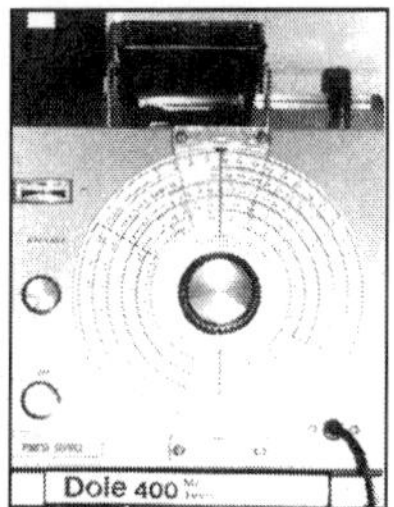

Fig. Dole Electric Seed Moisture Meter used in the USA.

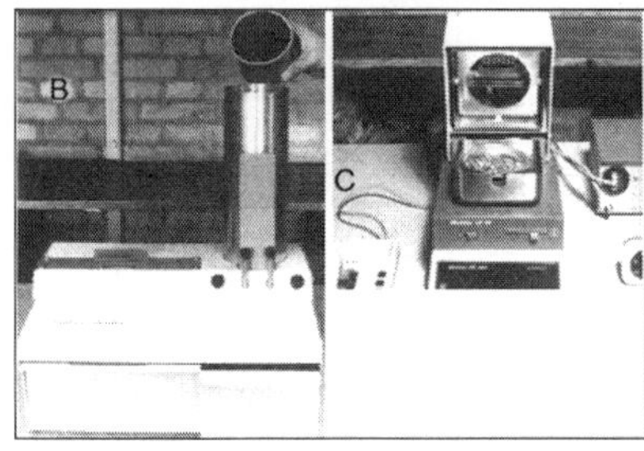

Fig. Electric Moisture Meters used in Denmark. A: Jacoby's infrared. B: Super-matic. C: Mettler.

TESTING MOISTURE CONTENT

The importance of moisture content in affecting the longevity of seeds in storage has been stressed in Chapter 7. In order to control the operations of drying (or moistening) of seeds in preparation for storage and to check the stability of moisture content during storage, it is clearly essential to have reliable methods of measuring the amount of moisture in a given sample.

Methods of determining moisture content of seeds have been classified by Justice into (a) basic methods in which the moisture is driven out of the seeds by heat and measured by the loss of weight of the original material, or the weight or volume of the condensed moisture, and (b) practical methods designed for rapid routine work and standardized against one or more of the basic methods. Probably all the moisture cannot be driven out of seeds without driving out small amounts of other volatile constituents or causing chemical changes in the material which would result in weight changes. In applying any method, therefore, it is necessary to adhere closely to the prescribed procedure in order that the results of all tests made by that method will be comparable.

Until recently ISTA prescribed three possible procedures:

(1) Drying in an oven for 17 hours at 103°C (2) Drying in an oven for one to four hours at 130°C and (3) Toluene distillation. Method (2) is applicable only to certain agricultural seeds and method (3), previously used for Abies, Cedrus, Fagus, Picea, Pinus and Tsuga, has now been eliminated because it had ceased to be used in practice (ISTA 1981 c). This leaves only method (1) - the "Low constant temperature oven method" as applicable to forest trees.

The test should be made on two samples of about 5 g each, drawn from the working sample including impurities, not on pure seeds. Large seeds should be ground, broken or cut into small fragments to facilitate drying and a good rule of thumb is that any seeds that average over 10 mm in diameter or length should be broken. The samples should be weighed and placed in metal containers, well-spaced to facilitate air circulation, within an oven which is maintained at a temperature of 103° ± 2°C for 17 ± 1 hours. At the end of that period the seed should be placed in a desiccator to cool for 30 – 45 minutes and then reweighed. The relative humidity in the laboratory where the final weighing is done should be less than 70 per cent, to avoid rapid re-absorption of moisture. The difference in MC of the two samples should not exceed a stated tolerance per cent. If it does, a further pair of samples should be tested; otherwise the mean of the two samples is the final result. The tolerance earlier prescribed by ISTA for all species was 0.2 per cent but, as pointed out by Gordon (1979) and Bonner (1981), a single tolerance figure is not applicable to all species. At the 1983 ISTA Congress in Ottawa, the tolerances approved for moisture tests in tree seeds were agreed as follows:

Sample Condition	Tolerance %
Small seeds, moisture <12%, *e.g.* Picea, Alnus	0.3
Large seeds, moisture <12%, *e.g.* Carya	0.4
Small seeds, moisture >12%	0.5
Large seeds, moisture 12 to 25%	0.8
Large Seeds, moisture > 25%, *e.g.* Quercus	2.5

For tropical tree seed laboratories wishing to conform to ISTA rules, this relaxation of tolerances will be a considerable help.

The calculation of moisture content should be made on a wet weight or fresh weight basis (see pp. 122–124) *i.e.*

$$\text{Moisture Content\%} = \frac{\text{Original weight} - \text{Oven weight}}{\text{Original weight}} \times 100$$

Although wet weight basis is prescribed by ISTA and is becoming increasingly the standard form for expressing moisture content, it is not yet universal. To avoid any doubt, the method of calculating moisture content should be stated explicitly on any certificate or statement of results.

As explained by Gordon and Rowe, provided that the initial fresh weight of a seed lot is measured and the initial moisture content (wet weight basis) calculated by oven-drying a sample, any new MC reached as a result of drying (or wetting) can be calculated directly from the new weight of the seed lot; there is no need for further oven-drying of samples at the new MC. The desired weight of the seed lot to be achieved through drying (or wetting) can be calculated by multiplying its initial weight by the initial dry matter percentage and dividing by the desired dry matter percentage.

e.g. (1) If initial wet weight of a seed lot = 50 kg and the MC (wet weight basis), determined by oven-drying a sample, is 25%, the oven-dry weight = 75% of wet weight = 37.5 kg.

(2) If a period of drying reduces the wet weight to 46.5 kg, the new MC

$$\left(\text{wet weight basis}\right) = \frac{(46.5-37.5)}{46.5} = \frac{9}{46.5} = 19.4\%$$

(3) If it is desired to reduce the MC (wet weight basis) to 10%, then desired oven-dry weight will be 90% of new wet weight and the seed lot must be further dried until its wet weight

$$\text{must be further dried unit its wet} = \frac{50\times 75}{90} = 41.67\text{kg}.$$

Electric moisture meters give rapid estimates of seed moisture but they are not considered accurate enough for official seed testing. Their rapid operation does make them very useful in certain situations, for example they should be accurate enough to check tree seed moisture as a guide to drying seeds for storage. Meter readings are converted to seed moisture content by means of charts supplied by the manufacturer or developed from calibration curves in the laboratory for the species in question. Most meters will not measure moisture above 15 – 20 per cent and require a minimum of 90 – 100 g of seeds for a test (Bonner 1981). A locally made, cheap and portable electric moisture meter has been used successfully in Thailand for several years to measure the MC of rice grains and could be used for tree seed of similar size. It measures electric capacitance and uses a 9-volt battery as the source of power.

Electric moisture meters are well suited to small seeds, but cannot be used for large seeds such as Juglans or Quercus; winged seeds such as Fraxinus are also difficult to measure. Large or winged seeds can be rapidly dried in a microwave oven. If the oven is preheated, drying can be completed in 5 minutes and weighing in 6 minutes immediately afterwards if in an electronic balance, or after 30 – 45 minutes' cooling in a desiccator if weighing is done in an ordinary balance. Results can be expected to be within 7 per cent at a probability of 0.05 in the case of large seeds of high MC such as Quercus, and within 2 per cent in Fraxinus and Carya, as compared with the more accurate results of the slower conventional methods.

A simple and cheap method of drying seeds quickly is to use an infra-red lamp (Gordon and Rowe 1982). A weighed sample is subjected to heat from an infra-red lamp with such an intensity that it loses all its moisture, without being burnt, in about 20 minutes. When the loss of weight ceases, the new weight is measured and the percentage loss calculated. An up-to-date account of the measurement of tree seed moisture content has been published recently.

OTHER TESTS

Other qualitative tests or observations may be made as the need arises, but do not call for detailed prescriptions. In many cases they can be combined with the purity test. They include:

AUTHENTICITY

There are several methods of determining whether the seeds are of the species stated.

They are:

i. Positive identification of the parent trees and their certification preferably on the basis of herbarium samples.
ii. Identification of the seeds by use of an analytical key or by comparison with a reference collection.
iii. Identification of the seedling. This may be the only way of determining if the seedlot is contaminated by hybrids or a mixture of two or more species with similar seed characteristics. A key and reference collection of seedlings will assist in identification.

Authenticating seeds as to provenance is not possible for most species, but some progress has been made in this field for Pseudotsuga andAbies and the use of isoenzyme techniques may open new possibilities.

DAMAGE, HEALTH

During the purity test, the operator should be alert to the incidence of mechanical damage and pathogenic infestation, which may indicate the need to improve the methods of transport or processing in use.

CALCULATION OF RESULTS

The following examples indicate the type of calculations required in the various stages of seed testing.

PURITY

Weight of full working sample	62.52 g
Weight of pure seed	56.89 g
$\text{Purity\%} = \frac{56.89}{62.52} \times 100$	= 91 %

SEED WEIGHT

The weight of 1000 seeds may be calculated as follows:

Either

(a) Seed weight determination carried out on 8 × 100 seeds from the pure seed component of the purity test.

Replicate No.	1	2	3	4	5	6	7	8	Total	Mean
Weight (g)	3.81	3.69	3.75	3.79	3.82	3.72	3.71	3.79	30.08	3.76

$$\text{Standard deviation} = \sqrt{\frac{n\left(\Sigma x^2\right)-\left(\Sigma x\right)^2}{n(n-1)}}$$

$$= \sqrt{\frac{8(113.118)-904.806}{8\times 7}} = 0.0496$$

$$\text{Coefficient of Variation} = \frac{0.0496}{3.76}\times = 1.32$$

Since this is considerably less than the maximum of 4.0 prescribed by ISTA, the sample is judged to be homogeneous and no further sampling is needed.

Weight of 1000 seeds = 3.76 × 10 = 37.6 g

Or

(b) Seed weight determination carried out on 1000 seeds from the pure seed component of the purity test, without replication.

Weight of 1000 seeds = 37.6 g

The number of seeds per unit weight may be derived as follows:

$$\text{Number of seeds per g of pure seed} = \frac{1000}{37.6} = 26.6$$

$$\text{Number of seeds per kg of pure seed} = \frac{1000\times 1000}{37.6} = 26{,}600$$

GERMINATION

Test made on 4 × 100 seed replicates from the pure seed component of the purity test.

Replicate No.	1	2	3	4	Total	Mean
No. germinated on completion of test	79	85	76	88	328	82
Sound seeds on cutting test	4	3	6	3	16	4

The range in number of germinated seeds from the largest to smallest of the replicates is 88 - 76 = 12.

Reference to Table shows that the maximum tolerated range for a mean germination of 82 per cent is 15. Since the actual range is less, the sample is accepted as homogeneous.

Germination % = 82%
Viability % = 82 + 4 = 86%

Viable Seeds per Unit Weight

A combination of the Viability % and weight of pure seed will give a figure for the number of viable seeds expected per unit weight of pure seed, while use of the germination % will indicate the number of germinable seeds.

Incorporation of a factor for purity % will express the numbers in terms of the number of viable or germinable seeds per unit weight of "impure" seed.

	Pure Seed		
		Per g	Per kg
No. of viable seeds	26.6 × 86 ÷ 100 = 22.9		22,900
No. of germinable seeds	26.6 × 82 ÷ 100 = 21.8		21,800
	Impure Seed		
		Per g	Per kg
No. of viable seeds	22.9 × 91 ÷ 100 = 20.8		20,800
No. of germinable seeds	21.8 × 91 ÷ 100 = 19.8		19,800

In the case of very small-seeded species for which the purity test is impracticable, the number of germinated seeds per unit weight of impure seed is determined directly by test. Figures for numbers of pure seeds per unit weight are not obtainable for this type of seed. Although usually expressed as "viable seeds per g", it should be noted that a cutting test is also impracticable for these small seeds, so strictly speaking the figures refer to germinable seeds. As an example:

Weight of replicate of impure seed (E. grandis) 0.10 g

Replicate No.	1	2	3	4	Total	Mean
No. germinated on completion of test	65	73	63	71	272	68

- No. of germinable ("viable") seeds per g = 680
- No. of germinable ("viable") seeds per kg = 680,000

Germination energy. Calculation of germination energy and energy period depends on the criterion used to define this. Table gives an actual example extracted from paul (1972). As mentioned above, the energy period may be arbitrarily defined in advance but is normally much less than the full period of the test. A single assessment is sufficient in this case. If, in the present example, the energy period had been defined as 12 days, then

$$\text{Germination energy} = \frac{29+31+44+36+28}{400} \times 100 = 41\%$$

Energy period = 12 days

If, on the other hand, the energy period is taken as that up to the day of peak germination, then daily assessment is necessary as shown in the table and

$$\text{Germination energy} = \frac{29+31+44}{400} \times 100 = 26\%$$

Energy period = 10 days

Table. Germination Test Sheet (extracted from Paul 1972)

Species: Pinus caribaea var. hondurensis Test No. 26/72
Seed Lot No. 85/71
Date shown: 2/11/11 Place: Mantin Nursery
Date completed: 30/11/11 Germination Per cent: 64%

Days after Sowing	Sub-Samples (4 × 100 seeds)				Daily Total	Cumulative Total	Cumulative Total as % of total seeds.	Mean daily germination %.	Daily Total as % of germinable seeds.	Cumulative total as % of germinable seeds
	A	B	C	D						
1		–	–	–	–	–	–	–	–	–
2	–	–	–	–	–	–	–	–	–	–
3	–	–	–	–	–	–	–	–	–	–
4	–	–	–	–	–	–	–	–	–	–
5	–	–	–	–	–	–	–	–	–	–
6	–	–	–	–	–	–	–	–	–	–
7	–	–	–	–	–	–	–	–	–	–
8	6	8	7	8	29	29	7.25	0.91	12	11
9	5	9	9	8	31	60	15.00	1.67	12	23
10	10	11	13	10	44	104	26.00	2.60	17	40
11	9	8	10	9	36	140	35.00	3.18	14	54
12	6	5	7	5	23	163	40.75	3.40	9	63
13	3	5	6	4	18	181	45.25	3.48	7	71
14	5	3	2	3	13	194	48.50	3.46	5	75
15	4	2	1	3	10	204	51.00	3.40	4	79
16	2	4	2	4	12	216	54.00	3.38	5	84
17	2	1	1	1	5	221	55.25	3.25	2	86
18	1	2	2	3	8	229	57.25	3.18	3	89
19	2	–	–	1	3	232	58.00	3.05	1	90
20	2	1	2	–	5	237	59.25	2.96	2	92
21	-	1	1	–	2	239	59.75	2.85	1	93
22	2	1	2	–	5	244	61.00	2.77	2	95
23	2	–	1	1	4	248	62.00	2.70	2	97
24	–	1	1	1	3	251	62.75	2.61	1	98
25	1	2	–	–	3	254	63.50	2.54	1	99
26	–	–	–	–	–	254	63.50	2.44	–	99
27	–	–	–	–	–	254	63.50	2.35	–	99
28	1	–	1	–	2	256	64.00	2.29	1	100
Totals	63	64	68	61	256				100	
Cut test	5	2	4	5	16					

$$\text{Mean germination}\% = \frac{63+64+68+61}{400} \times 100 = 64\%$$

$$\text{Viability}\% = \frac{256+16}{400} \times 100 = 68\%$$

Inspection of the pattern of germination suggests that rejection of all seeds germinating after peak germination would result in the rejection of an excessive proportion (60 per cent) of the potentially germinable seeds, while acceptance of all germinable seeds would unduly prolong the period of test and probably result in the inclusion of some seedlings of very poor vigour. A sensible rule of thumb applicable to the type of germination pattern in this example would be to define the energy period as lasting until daily germination falls to less than 25 per cent of peak. With this definition

$$\text{Germination energy} = \frac{29+31+44+36+23+18+13+10+12}{400} \times 100 = 54\%$$

Energy period = 16 days

Percentage of total germinable seeds which germinate within the energy period = 84 per cent.

Another commonsense measure of germination energy, used in Zimbabwe, is the percentage of germination when mean daily germination (cumulative germination divided by time elapsed since sowing date) reaches its peak. In the present example the peak of mean daily germination per cent is 3.48 per cent.

INDIRECT TESTS OF VIABILITY

Similar methods should be used as for germination tests *i.e.* 4 replicates should be used, the results tested for homogeneity and the mean number of apparently sound full seeds (cutting test) or stained embryos (tetrazolium test) expressed as a percentage of the total pure seeds tested.

4

Seed Saving

INTRODUCTION

Saving seed from your own vegetable garden to plant next year is a great way to save money. But it's also a way to preserve heirloom and open-pollinated varieties of plants and thus contribute to the stability of both your personal and our collective food system.

Plants that reproduce through natural means tend to adapt to local soil and climate conditions over time, and evolve as reliable performers in that locale. They will naturally become more disease and insect resistant over the years.

However, the messiness (and un-profitability) of the natural way has led corporate-funded scientists to develop cloned and otherwise genetically modified seeds. In the past, selective breeding has led to plants that are more disease resistant and hardy. But corporations have spun out of control in recent years, creating seeds that are reliant on chemicals that they also produce, and removing or adding genetic material in ways that are impossible in Nature.

In order to protect their investment in researching and producing these seeds, governments have allowed corporations to patent the varieties they create–and prevent them from being saved. In this manner, thousands of varieties of vegetables and flowers are being lost, eroding the gene pool and actually resulting in less hardy, more vulnerable plants. The process is also putting a few large corporations in charge of our food supply, with a lot of experimentation and little independent testing or consultation! In an attempt to safeguard their food supplies, many countries have, in the past, set up seed banks. Journalist Fred Pearce notes that seed saving is crucial to the future of plant breeding: "The genes found in millions of crop varieties contain vital traits that plant breeders need to improve modern varieties, raising yield, protecting them against new diseases, and adapting them to climate change."

Seed saving and swapping also provide a great science lesson for children… helping them learn more about where their food comes from and how it's produced, as well as about the cycle of life. Very young kids can begin by growing a few gourd vines. Simply allow them to dry out and store in a dry place. Next Spring, simply open the gourd and shake out the seeds. They can then progress to collecting other vegetable and flower seeds, and sharing them with other gardeners. So seed saving and sharing is truly an activity for anyone who really cares about their food and wants to protect its diversity. Caring for a self-perpetuating garden is fun and rewarding. And swapping seeds with other gardeners at your local seed swap can be a great social experience. Saving seeds from your garden isn't difficult. And you can begin right now. The first step is to be sure you're planting open-pollinated varieties rather than hybrids. As the plants grow, select a few healthy specimens from which you will harvest the seeds.

PROCESS OF SEED SAVING

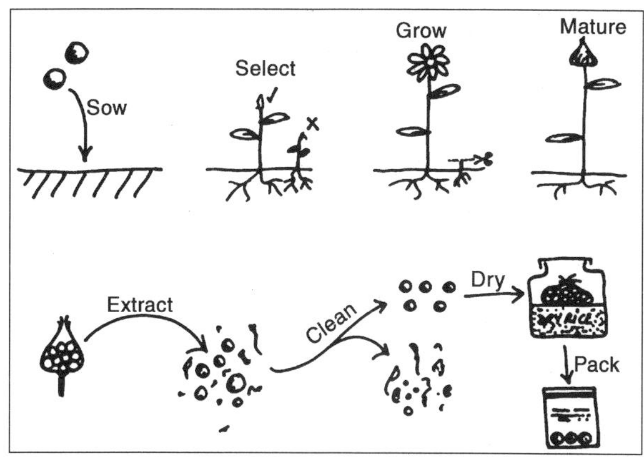

BROAD BEANS

Broad beans will cross with other varieties that are growing nearby. So if you want to keep your variety pure, you need to isolate them in some way. Theoretically you should aim for at least half a mile between varieties. In practice, in a built up area, fences, trees and houses will all reduce insect flight. This means you should have minimal crossing even with beans much closer than half a mile so long as none of your immediate neighbours are growing different varieties of bean.

In an open situation like an allotment, you can physically isolate plants. Broad bean pollen is transferred by insects working the flowers, but the plants will also self pollinate, so if you can exclude insects at flowering time, say by a covering of fleece, your seed crop will be pure.

The simplest method of all, if you are growing a relatively large number of beans and you are not concerned about achieving 100 per cent purity (eg just for your own use), is to mark and save seed from several plants in the middle of a block of beans. Insects are relatively unlikely to come from a neighbouring patch straight to the middle of your patch, tending to work the outside flowers first. So by the time they reach your seed beans, the amount of 'foreign' pollen remaining should be small. Always keep seed of strong, healthy plants and get rid of any that are not typical of the variety ideally before they flower.

Let your seed beans mature and dry on the bush. The pods will turn dark drown, dry and wrinkled. Then pick and shell them out. Check that they are really dry by biting on them. If your teeth leave a dent, dry them further in a warm (not hot) place with a good flow of air. Broad bean seeds should keep for several years, so there is no need to grow plants for seed every year.

FRENCH AND RUNNER BEANS

It is important to grow some bean plants specifically for seed, rather than simply collecting the left-over pods at the end of the season. The plants should be good strong specimens, and any that are less healthy looking or not true to type for the variety should not be used for seed production.

French beans are self-pollinating, mostly before the flowers open. Despite this, they can be crossed by insects with other varieties nearby. The extent of crossing varies by area. If you are just saving seed for your own use, grow your seed crop of french beans at least 6 feet away from any other variety (12 feet if possible), and you are unlikely to have a significant problem with crossing in the UK.

Runner bean flowers need to be 'tripped' by wind or insects before the beans set, and are much more likely to cross with other varieties grown nearby than french beans. Ideally, to be sure that no crossing takes place, seed crops of runner bean should be at least 1/2 a mile away from any other varieties of runner bean. Bear in mind, though, that buildings, trees, and other barriers will limit insect flight patterns, and if you are gardening in a town or built up area, you are likely to have relatively little problems with crossing unless your immediate neighbours are also growing runner beans. If they are - or on an open site such as an allotment—your only answer may be to try to persuade your neighbours to grow the same type of runner.

To collect the seeds, allow the pods to mature fully on the plant until they start to yellow and dry out. In wet weather, collect the pods individually as they get to this stage. Then spread out somewhere out of the rain with a good airflow until the pods are fully dry and brittle. Once they are dry, shell out the beans and dry further out of the pods. The beans should be dry enough that they break when you bite on them, rather than leaving a dent. Store in an airtight container. If they are well dried, and stored in a cool dark place, the beans will last around 3 years.

If you have problems with weevils eating your seeds, put the sealed container in the freezer for a week immediately after drying the beans; this will kill any insect eggs before they hatch. When you take them out, let the container come up to room temperature before opening it, otherwise the beans will absorb moisture from the air.

PEAS

Peas are almost entirely self pollinating, only very occasionally crossing with other plants. Set aside a section of row that is entirely for seed production, and make sure you sow at a time that will avoid pea moth To avoid physical mixing up of the seeds, separate different varieties of pea with another crop. Check the row from time to time as the peas grow, and pull up any plants that are weak or not true to type.

Let the peas mature until the pods are brown and the seeds start to rattle. If the weather is very bad, pull up the whole plants and bring inside (for example hung upside down from the shed roof) once the pods start to wither, to ripen and dry further. Once the pods are really dry, shell the peas out. Dry the shelled peas further in a warm (but not hot) place, label with the variety and date, and store.

AUBERGINES

Aubergine flowers are mainly self pollinated, but can be crossed by insects. So if you are planning to save seed, you should only grow one variety. Aim for 6 to 8 plants each year to maintain a variety long term. For 100 per cent isolation you need 50 feet between your seed plants and any other aubergines. If you are growing them in a greenhouse/polytunnel you should be able to get away with a somewhat smaller distance.

To get ripe seeds let the fruits mature well past eating stage. Purple/black cvs turn a muddy purple-brown colour, green/white cvs turn yellowish. Mark 1 or 2 early good fruits on each plant to leave for seed, and then pick and eat later fruits.

To remove the seed, cut into quarters lengthwise, avoiding the core, and pull apart. The hard brown seeds should be obvious. Put the quarters into a bowl of tepid water, and rub the seeds out with your fingers. You may need to pull them apart to get all of the seeds. Add more water, stir thoroughly, and wait a few minutes. Good seeds will sink to the bottom, leaving debris and poor quality seeds on the surface. Pour the debris off gently through a sieve, then refill with water and repeat a couple more times. Eventually you will be left with good seeds in plain water. Empty into a clean sieve, shake to remove as much water as possible, and then tip on to a plate and spread out well. Put to dry somewhere warm but not hot, and mix occasionally to make sure that they dry evenly and don't stick together. Aubergine seeds will keep up to 7 years if dried thoroughly and stored in a cool dark place.

SWEET PEPPERS AND CHILLIES

Sweet peppers and chillies are both members of the same species, Capsicum annuum (some less common chillies come from other capsicum species).

Pepper flowers are self pollinating, and will set fruit without any insect activity. However, they will also cross readily, and sweet peppers will happily cross with chillies. You need to isolate your plants by around 150 feet (50 metres) from any other peppers or chillies growing nearby. Even if you are only growing one variety be careful about other varieties growing in adjacent gardens or allotments.

If you want to grow several varieties, or if your near neighbours are also growing peppers, you could consider making an isolation cage to cover 3 or 4 plants. This is easy to do, and costs very little, especially if you can get hold of some old net curtain material. You can put a cage up over plants grown in pots, growbags or directly in the ground.

To save the seed, take peppers on your isolated plants which have ripened fully to their final colour (usually yellow or red). Cut the peppers open carefully, and rub the seeds gently off of the 'core' onto a plate. Wear rubber gloves to deseed chillies, as the chilli oil sticks to your fingers and is very hard to wash off. Dry the seeds in a warm but not hot place until they snap rather than bending

MAKING AN ISOLATION CAGE

To make a simple isolation cage ideal for peppers or aubergines, you need some cheap nylon flyscreen 5 times as long as it is wide, four canes or thin stakes, and some string and garden wire. Alternatively, you can use old net curtains, or other netting small enough to exclude insects. A piece of screen 1m by 5m will give a cage large enough to cover 3 or 4 plants.

Cut a square piece of screen 1m x 1m to make the top of the cage, and then fold the remaining strip of flyscreen round and sew its ends together. The resulting band will be the sides of the cage. Then sew the top to the sides, making a cube of flyscreen with the bottom missing.

To put up the cage over your plants, hammer the four canes into the ground in a square a little smaller than the cage top, so that they stick up a little less than the height of the cage. Twist a short piece of wire tightly round the top of each cane, and then run string in a square around the tops of the canes, supported by the wires to stop it slipping. Run a second piece of string around the stakes lower down to stop the sides of the cage blowing in against the plants. Then slip the cage over your plants, and weigh it down with earth or rocks.

TOMATOES

Most modern varieties of tomato are self pollinating, and will not cross. The anthers on tomato flowers (which make the pollen) are fused together to make a tight cone that insects cannot enter. Usually the stigma (the receptive surface for receiving pollen) is very short, and so is located deep inside this cone of anthers. No insects can get to it and the only pollen that can fertilise it comes from the surrounding cone of anthers.

In a few varieties however, the stigma is much longer, sticking out beyond the cone of anthers. In this case, insects can get to it, and there is the chance of cross-pollination. Varieties with longer stigmas include potato leaved tomatoes and currant tomatoes. To avoid crossing only grow one variety with exposed stigmas. The double flowers which are sometimes formed first by many beefsteak tomatoes also often have exposed stigmas, but later single flowers will be normal. To collect the seed, allow your tomatoes to ripen fully. Then collect a few of each variety that you want to save seed from. Slice them in half across the middle of the fruit, and squeeze the seeds and juice into a jar. You then need to ferment this mixture for a few days - this removes the jelly-like coating on each seed, and also kills off many diseases that can be carried on the seeds. To do this put the jar of seeds and juice in a reasonably warm place for 3 days, stirring the mixture twice a day. It should develop a coating of mould, and start to smell really nasty!

After 3 days, add plenty of water to the jar, and stir well. The good seeds should sink to the bottom of the jar. Gently pour off the top layer of mould and any seeds that float. Then empty the good seeds into a sieve and wash them thoroughly under running water. Shake off as much water as possible, and tip the sieve out onto a china or glass plate (the seeds tend to stick to anything else). Dry somewhere warm but not too hot, and out of direct sunlight. Once they are completely dry, rub them off the plate and store in a cool dry place, where they should keep well for at least 4 years.

BEETROOT, CHARD AND LEAF BEET

Beetroot, leaf beet/perpetual spinach, swiss chard and sugar beet are all members of the same family and will cross readily. They are biennial, and flower in their second year. Chard/leaf beet for seed are overwintered in situ, and will be fine in most of the UK. Select a minimum of six to eight plants to leave for seed which best fit your needs (depending on your preference for stem versus leaf, smooth or wrinkled leaves etc). Beetroot can also be overwintered in situ, or can be harvested in autumn, the best plants selected and stored then replanted in spring.

All types of beet will cross with one another, and since the flowers are wind pollinated, crossing can take place with any other flowering beet plants within around 2 miles. How fussy you need to be about crossing depends on what you are trying to achieve. If you simply want a reasonably diverse population of leaf beet, a degree of crossing is not that important. Plant your seed plants closely together in a square, and take seed from the central plants in the block; you will find that the amount of 'contamination' is minimal providing there aren't large numbers of other flowering beets right next door.

If you are aiming to keep a variety true to type you need to isolate it, usually by physically covering your seed plants. To do this, plant at least six plants very close together in a circle, with a wooden stake in the middle. As the seed stalks form, growing up to four feet tall, tie them together, supported by the stake. Then as they develop cover the group of flower heads with either a shiny paper bag that will withstand rain, or a bag made out of agricultural fleece. Shake the bag from time to time to make sure that pollen is distributed within the bag.

As the large, prickly seeds mature, keep an eye on them, and start to harvest as they turn brown and start to dry out. You can either cut entire seedstalks, or harvest mature seeds by rubbing them into a bucket. Make sure that the seeds are thoroughly dry before storage, and they should last at least five years.

CARROTS

Carrots are biennial, flowering in their second year of growth. In areas with mild winters, leave your carrots in the ground, mulching them heavily. The foliage will die back in autumn, but will then resprout and start to flower in the spring. In colder areas, dig up your carrots in the autumn, and select the best coloured and shaped roots. Twist off the foliage, and store the roots in a box of dry sand in a frost free place, making sure that they don't touch. In spring, replant the roots, and they will resprout and flower.

If you want to maintain a carrot variety effectively, you really need to save seed from at least 40 good roots to maintain good genetic diversity. If you have too small a genetic pool, you will end up with small, poor quality roots in a very few generations.

Carrots grow into big plants waist high or taller, producing successive branches with large flat umbels of flowers. They are insect pollinated, and need to be isolated from other flowering carrot varieties by at least 500m in an open field situation. This is not normally a big problem, since few people let their carrots go to seed. However, they will cross with wild carrot (Queen Anne's Lace), giving thin white useless roots. As with all insect pollinated crops, barriers such as houses, tall hedges and other high crops can affect insect flight paths drastically, so you don't necessarily need to eliminate all Queen Anne's Lace within a 1/2 km radius; but do watch out for any white roots in subsequent generations and get rid of them.

To harvest your carrot seed, keep an eye on the umbels of flowers, and cut them off with secateurs as they start to turn brown and dry. If you have plenty of plants, just save seed from the first and second umbels of flowers to appear on each plant, as these will give the biggest and best seed. Dry the seed heads further inside, and then rub them between your hands or in a sieve to separate them. You will notice that the seeds have a 'beard' which is removed in commercial seed to make them easier to pack. You can sieve the seeds further to remove more of the chaff, but there is no need to get the seed completely clean - just sow slightly more thickly to allow for the chaff mixed in. Carrot seed is relatively short lived, but if it is stored somewhere cool and dry, it should give good germination for 3 years.

HERBS

Basil, coriander and dill are annuals, parsley is a biennial, flowering in its second year of growth. Basil flowers are insect pollinated, and different varieties flowering within around 150' of one another may cross. On a garden scale, if you want to grow several types of basil, just keep picking the flower stalks off of all the varieties apart from the one that you want to grow for seed. Once several flower spikes have set and the flowers have started to wither, mark those spikes for saving seed from, and you can then allow the other varieties to flower. The seeds are ready to collect when the spikes turn brown and dry out. Don't worry about the seeds dropping out - they are well attached, and actually need quite a lot of rubbing to free from the dead flower heads.

With both coriander and dill, to get the best seed for sowing in future years, pull up and discard the earliest plants to bolt, and only save seed from those plants that produce plenty of leaf and flower late. It is best to plan to save seed from early summer sowings, to allow plenty of time for the seed to mature and dry on the plant. Harvest as soon as the seed is brown and dry, as it does tend to drop from the seed heads. Rub the heads together in your hands over a bucket to free the seed. Dill seed usually comes cleanly away from the seed heads. Coriander seed tends to contain more chaff, but you can winnow it by pouring gently from one bucket to another in a light breeze if you want to clean it for kitchen use.

To save parsley seed, overwinter at least two or three plants. In warmer areas mulch heavily with straw or cover plants with a frame, elsewhere grow a few plants in a polytunnel or greenhouse. The next spring, the plants will start to flower and produce seed. Flat and curly leaved varieties will cross, as the flowers are insect pollinated, so you should only grow one type for seed at a time. Harvest the seeds from individual flowerheads as they dry and turn brown, as they tend to drop from the plant when ready.

BROCCOLI, KALE AND CABBAGES

Sprouting broccoli, cabbages, cauliflowers, calabrese, kales and brussels sprouts are all members of the same family (Brassica oleraceae), and will all cross with each other. They won't cross with turnips, swedes, oriental brassicas or mustard greens. In addition, they are mainly self-incompatible—which means that in order to get seed, insects have to carry pollen from one plant to another to pollinate the flowers. Because of this, you can't simply grow your broccoli or cabbages for seed in an insect proof cage to avoid crossing. So long as you only seedsave from one member of the family in any given year, you can grow as many other brassicas as you like without problems so long as you don't let them flower. For absolute seed purity, make sure that there are no other flowering brassicas within a mile of your garden. In practice, fences, trees and tall crops all break up insect flight patterns, so as long as you don't have any immediate neighbours with flowering crops in their garden, you shouldn't have too many problems with crossing. To make it as easy as possible for insects to work your seed plants, make sure that they are laid out in a block, rather than a row, so that bees tend to move from one plant to another, rather than away to other flowers elsewhere.

Keep at least six plants for seed, ideally more. Remove any poor specimens, or any that are not typical for the variety -you can always eat these plants, so long as you don't allow any flowers to open.

All of the brassicas, including cabbages, will throw up a tall flower stalk covered in lots of small yellow flowers. These will then form slender seed pods, which start out green, and turn a straw colour as they mature and dry. Once they start to dry, keep a close eye on them, as they tend to shatter and drop their seed. Its best to cut entire plants once most of the pods begin to look dry, and then leave them to mature further on a sheet indoors. Once they are thoroughly dry, the seeds will come out of the pods very easily; the simplest way is to trample the plants on top of a large sheet, and then sieve out the debris.

TURNIPS AND THE ORIENTAL BRASSICAS

Mizuna, pak choi, tatsoi and mibuna are all sub varieties of Brassica rapa—the same family as turnip. This means that although they will cross with each other, or with turnips in flower, they won't cross with broccoli or cauliflowers. Although you can only grow one of these vegetables for seed in any year, you can of course grow any of the others for kitchen use, so long as you don't allow them to flower at the same time as your seed plants.

To grow an oriental brassica or turnip variety for seed, you usually need to overwinter the plants. They are naturally biennials, producing their flowers and seeds in their second year of growth. Although spring sown crops may bolt to seed in hot summer weather, this is not ideal for seedsaving, as you may end up accidentally selecting for early bolting in future years. The best solution is to sow your seed crop after midsummer in a polytunnel, where semi-mature plants will overwinter quite happily in all but the coldest parts of Britain. If necessary you can give extra protection in cold weather by putting fleece over plants inside the tunnel.

Select at least 6 of the healthiest and most typical plants to reserve for seed, eating the rest over the winter. In spring, the plants will flower, and then form seedpods. Make sure that there is good insect access to the tunnel at this point so that the flowers are pollinated.

The seedpods are green at first, but then gradually dry out and turn a pale tan colour. Once most of the pods are dry and brittle, cut the entire stalks of the plant, and lay out on a sheet somewhere undercover with a good airflow to finish drying off. Then rub and crush the pods with your hands to release the seeds, and separate the seeds from the chaff with a coarse sieve.

LETTUCE

Lettuce flowers are self pollinating, and very rarely cross. If you plan to save seed from more than one variety of lettuce, separate them by around 12 foot or plant a tall crop in between the rows.

Select two or three good lettuces from your row, and mark them for seed. It is very important not to save seed from any plants that bolt early, as you want to select for lettuces that stand well. Heading lettuces may need a little help for the flowering stalk to emerge; slitting the heads partially open with a knife works well.

Once the lettuces have flowered, the seeds will ripen gradually, starting in about a fortnight. Harvest seed daily to get the maximum yield, shaking into a bag. Or wait until a reasonable number of seeds are ready and then cut the whole plant. Put it head first into a bucket, shaking and rubbing to remove the seeds. If you leave the whole cut plant upside down in the bucket somewhere dry, slightly immature seeds will continue to ripen over the next few days.

Most of what you have collected in the bucket will be white 'feathers' and chaff. To sort the seed, shake it gently in a kitchen sieve. Some seeds will fall through the sieve, with the rest collecting in the bottom. The feathers and chaff will rise to the top, and you can pick them off. There's no need to get the seed completely clean; a little chaff stored and planted along with the seeds won't cause any harm.

If the seed feels a little damp, dry it further on a plate before labelling and storing. Lettuce seed should keep for around 3 years, provided it is kept cool and dry.

PUMPKINS, COURGETTES, MARROWS AND SQUASHES

Beware that pumpkins, squashes, marrows and courgettes will all cross readily with each other. The best (usually only) way to save pure seed on a home scale is to hand pollinate one or more fruits. This is very easy and will avoid disappointments with lumpen squash/courgette crosses. The explanation given here is for pumpkins, but applies equally to squashes, courgettes and marrows.

Pumpkin plants have two different types of flower, male and female. The female flowers are the ones that will grow into pumpkins. They can be identified by the small immature fruit which should be obvious beneath the flower. Male flowers just have a straight stem. You need to transfer pollen from a male flower into a female flower, making sure that no pollen gets introduced from plants of a different variety.

One evening, when the plants are just beginning to produce flowers, find some male and female flowers that are going to open the next day. Buds that are just ready to open are much fatter than the others, and they have turned from green to yellow.

You need to stop these flowers opening, so that insects can't get into them. The easiest way to do this is to gently slip a thin rubber band over the end of the petals, to hold them shut.

The next morning go back to the plants. Pick a male flower, take off its rubber band, and tear off the petals. Gently take the rubber band off of one of your female flowers. Using the male flower like a brush, rub the pollen on to each section of the stigma in the centre of the female flower.

Then carefully rubber band the female flower shut again so that no insects can get in with more, 'foreign', pollen. Tie a piece of wool loosely around the stem of the female flower, so that at harvest time, you know which pumpkins you have hand pollinated.

Now leave the pumpkins to develop and ripen. After you have harvested them, keep them in a cool dry place for another month or so to ripen further indoors.

Then cut the pumpkin in half, and scoop out the seeds, leaving the rest of the fruit for cooking as normal. Wash the seed in a colander, rubbing it between your hands to get rid of the fibres, and then shake off as much water as possible.

Spread the seed out on a plate to dry. It needs to dry as quickly as possible, but without getting too hot, for example on a sunny windowsill. To test whether the seeds are dry enough, try bending one in half. If it is dry, it will snap rather than bending.

MELONS AND CUCUMBERS

All varieties of melon will cross. Ideally, you need around a quarter of a mile between different varieties. If your melons are in a greenhouse or tunnel, you can probably get away with a somewhat smaller distance, particularly if there are hedges, houses or other tall barriers in between your melons and the neighbouring crop. Cucumbers won't cross with melons, but will cross with any other cucumbers or gherkins nearby. Again, you need around a quarter mile isolation to make sure that your plants won't cross.

It is possible, although fiddly, to hand pollinate both melons and cucumber flowers. Grow plants under a fleece tunnel to exclude insects, and then hand pollinate the flowers on those plants with a paintbrush. Make sure that you exchange pollen between different plants to keep the diversity of your variety. To harvest melon seed, pick the melons when they are ripe and ready for eating and keep indoors for a further day or two for the seed to mature further. Then open the fruit, scoop the seed out, and wash in a sieve under running water. Spread out on a china plate to dry thoroughly.

Cucumbers need to be ripened well beyond the edible stage. They will become much fatter, and green varieties will turn a dark yellow brownish colour, white varieties a paler yellow. Keep for a week or so after picking to let the seeds mature fully. Then cut open, scoop out the seeds and surrounding pulp into a jamjar, add a little water and stir well. Leave the jar on a sunny windowsill for 2-3 days for the seeds to ferment. On the third day, fill the jar fully with water, and stir well again. The good seeds should sink to the bottom of the jar, leaving pulp, debris and empty seeds floating on top. Gently pour off the water and debris, refill the jar, and repeat. After a couple of rinses, you should be left with good seeds at the bottom of a jar in clean water. Drain off the water, and spread out on a plate to dry well.

SAVING TOMATO SEED

One can start collecting viable seeds for saving once the fruit reaches the "breaker" stage. This is when a small amount of colour (usually red) shows up at the blossom end of the fruit. Still it is probably best to wait till one a fruit is fully coloured. Selecting over-ripe fruit can result in reduced seed vigour and germination even though this is part of nature's method of removing the seed's gelatinous coat which contains germination inhibiting compounds. Over-ripe fruits can also result in precious seed germination in some cultivars (means the seed germinations in the fruit). Trying to speed up the ripening process up by utilizing ethylene (the gas which speeds ripening) can also result in reduced vigour and germination.

Fermenting the seeds will help to remove the gelatinous coat which contain gemrination inhibitors and helps reduce or destroy any pathogens present. Though messy/smelly, I recommend it. Extract the seeds and put them in a marked container. Squeeze juice and seeds into 8oz-16oz plastic containers with lids (which will help too keep out insects, prevent spills and hold moisture) for about 2 days (it is also wise to label the container rather than the lid to prevent any mix ups). If one does use coverings, the container should not be set in direct sunlight. The warmer the temps, the quicker they ferment. 3 days or more may be needed in cool conditions. Rely on the fruit juices for moisture but add a small amount of water if needed to prevent drying. Too much water and the may start sprouting.

A mold should start forming on the surface but if it does not and its been 2 days that's ok. Just rinse the seeds well in a strainer and them tamp then out to dry. One can do so on paper towel, a plate, a screen or various other methods. It is important to dry seeds slowly, since rapid drying out shrinks the seed coat around the embryo and reduces seed quality (in terms of vigour). Good seed should be a light tan colour. If you get dark brown seeds they may have set too long and germination will likely be reduced. If this happens do not worry too much, usually some will germinate the next year.

Store the seeds in the refrigerator and they should keep well for about 4 years. One can still get viable seeds 7–12 years later but usually the germination percentage is lower. If one plans to save seed this long be sure to save plently of seed. If one reduces the humidity this will increase seed life. This is done by adding a desiccant like silica to a container which will hold the packets of seeds. One can get desiccants from the packets of silica that come packed in electronics.

If one suspects diseases treat seed in a 10 per cent clorox solution for 10-20 minutes, then rinse, just PRIOR to sowing (NOT storage). This can reduce germination somewhat but can help to eliminate disease problems that may be associated with the seed. It cannot do anything for virus diseases however.

SAVING SEED FROM THE GARDEN

Every year a few gardeners ask about saving seed from their flowers and vegetables. We would not have the wonderful heirloom varieties if someone hadn't kept the seeds year to year. Seed saving can be a rewarding and cost saving way to garden, but beware of the pitfalls. Not every plant's seeds are worth keeping. Hybrid plants are developed by crossing specific parent plants. Hybrids are wonderful plants but the seed is often sterile or does not reproduce true to the parent plant. Therefore, never save the seed from hybrids. Another major problem is some plants' flowers are open pollinated by insects, wind or people. These plants include squash, cucumbers, melon, parsley, cabbage, chard, broccoli, mustard greens, celery, spinach, cauliflower, kale, radish, beets, onion, and basil. These plants cross with others within their family. The only way to maintain the original variety is to isolate by large distances. Isolation is often impossible or impractical in a home garden.

Some seeds may transmit certain diseases. A disease that infected a crop at the end of the growing season may do little damage to that crop. However, if the seed is saved and planted the following year, the disease may severely injure or even kill the young plants.

What can you save? Standard or heirloom varieties that are not cross-pollinated by nearby plants are good candidates. Many gardeners successfully keep beans, tomatoes, lettuce, and peppers. Plants you know are heirloom varieties are easy to save. Ask the person or organization you obtained the seed from how they did it. Some people like to experiment, but make sure you don't bet the whole garden on saved seed.

When saving seed, always harvest from the best. Choose disease-free plants with qualities you desire. Look for the most flavourful vegetables or beautiful flowers. Consider size, harvest time and other characteristics. Always harvest mature seed. For example, cucumber seeds at the eating stage are not ripe and will not germinate if saved. You must allow the fruit and seed to fully mature. Because seed set reduces the vigour of the plant and discourages further fruit production, wait until near the end of the season to save fruit for seed.

Seeds are mature or ripe when flowers are faded and dry or have puffy tops. Plants with pods, like beans, are ready when the pods are brown and dry. When seeds are ripe they usually turn from white to cream coloured or light brown to dark brown. Collect the seed or fruits when most of the seed is ripe. Do not wait for everything to mature because you may lose most of the seed to birds or animals.

Beans, peas, onions, carrots, corn, most flowers and herb seeds are prepared by a dry method. Allow the seed to mature and dry as long as possible on the plant. Complete the drying process by spreading on a screen in a single layer in a well-ventilated dry location. As the seed dries the chaff or pods can be removed or blown gently away. An alternative method for extremely small or lightweight seed is putting the dry seed heads into paper bags that will catch the seed as it falls out.

Seed contained in fleshy fruits should be cleaned using the wet method. Tomatoes, melons, squash, cucumber and roses are prepared this way. Scoop the seed masses out of the fruit or lightly crush fruits. Put the seed mass and a small amount of warm water in a bucket or jar. Let the mix ferment for two to four days. Stir daily. The fermentation process kills viruses and separates the good seed from the bad seed and fruit pulp. After two to four days, the good viable seeds will sink to the bottom of the container while the pulp and bad seed float. Pour off the pulp, water, bad seed and mold. Spread the good seed on a screen or paper towel to dry.

Seeds must be stored dry. Place in glass jar or envelopes. Make sure you label all the containers or packages with the seed type or variety, and date. Put in the freezer for two days to kill pests. Then store in a cool dry location like a refrigerator. Seed that molds was not sufficiently dry before storage.

Seed viability decreases over time. Parsley, onion, and sweet corn must be used the next year. Most seed should be used within three years.

TIPS

Seed saving is not always feasible with all types of vegetables, but collecting your own seed can be an exercise in self-sufficiency and a lesson in plant biology. Seeds you save from your home production system are accustomed to your climate and growing medium and are adapted to pests in your area. Seeds are generally saved from annual and biennial plants. Perennials are usually propagated through division or cuttings.

The easiest seeds to save are open-pollinating, non-hybrid annuals. Plants that are not self-pollinating can cross-pollinate; therefore, it is best to grow only one variety of a plant from which you want to save seed that season. If two varieties of spinach bloom near each other, the resultant seed is likely to be a cross between the two. Different varieties of peppers should be separated by 500 feet to avoid cross-pollination. Melons, pumpkins, cucumbers, and squash need even more personal space—at least a half-mile is required.

Biennials require more work and commitment. These plants do not send up seed stalks until the second season. Biennials include beets, Brussels sprouts, cabbage, carrots, cauliflower, celery, onions, parsley, parsnips, rutabaga, salsify, Swiss chard, and turnips.

Do not save seed from hybrid varieties if you want plants like the parents. Seeds from hybrid varieties produce a mix of offspring, many of which may have different characteristics than the parent. Seed from hybrid vine crops is often quite variable also—squashes, cucumbers, melons and pumpkins often cross-pollinate with other genetically compatible varieties. Unless pollination has been strictly controlled, strange hybrids often result in the next generation.

Among the vegetable seeds most easily saved are non-hybrid tomato, pepper, bean, eggplant, cucumbers, summer squash, and watermelons. Collect seeds from the fully mature, ripe fruit of these plants.

TOMATO

The seeds are encased in a gelatinous coating, which prevents them from sprouting inside the tomato. Remove this coating by fermenting it. This mimics the natural rotting of the fruit and has the added bonus of killing seed borne tomato disease. Squeeze the seeds from a fully ripe fruit into a bowl, add water and let stand at room temperature for about three days. Once fermentation occurs, mold will form on the surface of the water. Add more water, stir, then gently scrape mold and debris off the top. Repeat until only clean seed remains, strain, rinse, and leave the seeds at room temperature until they are thoroughly dry.

PEPPERS

Select a mature pepper, preferably one that is completely red. Cut the pepper open, scrape the seeds onto a plate and let the seeds dry in a non-humid, shaded place, testing them occasionally until they break rather than bend. Leave at room temperature until completely dry.

BEANS, PEAS, AND OTHER LEGUMES

Leave pods on the plant until they are "rattle dry." Pick the pods and remove the seeds when completely dry.

EGGPLANT

Leave the plant on the vine until it is well past the stage when you would pick it for kitchen purposes. Eggplants ready for seed saving will be dull, off-coloured and hard. Cut the eggplant in half and pull the flesh away from the seeded area.

CUCUMBERS

Cucumbers change colour after they ripen and start to become mushy. Cut it in half and scrape the seeds into a bowl. Remove their slimy coating by rubbing them gently around the inside of a sieve while washing them or soak them in water for two days. Rinse and dry.

SUMMER SQUASH

Summer squash is at the seed-saving stage when you cannot dent the squash with a fingernail. Cut it open, and scrape the seeds into a bowl, wash, drain, and dry.

WATERMELON

Put the seeds from ripe fruit in a strainer and add a drop of dishwashing liquid to remove any sugar from the seeds.

STORING SEEDS

Store most seed packets in airtight jars. The exception is legumes, which store best in breathable bags. To keep the seeds dry, fill a small cloth bag with about one-half cup dried powdered milk. Place the packet in the jar beneath the seed packets. Be sure to label your container with the variety, the date, and other pertinent information. Store your seeds in a cool, dark, dry place; a refrigerator is a good choice. Avoid opening the container until you are ready to plant.

Stored seeds will retain their viability for different lengths of time depending on the type of seed. Melon seed can be stored for as long as five years, while sweet corn is only good for one year. Other types of seed remain viable for two to three years.

5

Seed Treatment

TECHNOLOGY

Successful seed treatment is a complex story as it depends on a number of interacting factors. Full compliance with all of them is essential to achieve a high quality seed treatment. It all starts with the seed being treated needing to be clean and of high quality itself.

The seed treatment formulation has to be stable and have basic adhesive properties; the film coating has to complete the adhesiveness in order to ensure even distribution of the seed treatment product on the seeds and to provide good seed flowability. The slurry recipe has to be adapted to the seeds, and the volume needs to be adjusted to achieve good coverage and a homogeneous distribution among the seeds. Finally, the equipment must be able to be adjusted precisely and has to work reliably. The treatment will be of high quality only if the operator applies exactly the right dose of both seed treatment product and film coating.

Seed treatment technology has come a long way since the use of salt brine in the mid-1600s. Today, seed treatments deliver clear environmental, economic and social benefits, making the technology a perfect tool for sustainable agriculture.

With a growing world population, now more than ever it's critical that farmers have access to the tools that will help them grow more food while protecting the environment. And seed treatments are one of these tools.

"By utilizing modern agriculture technologies, farmers will be able to boost yields, conserve water usage and protect biodiversity," says Keith Jones, director of stewardship and sustainable agriculture for CropLife International based in Brussels, Belgium. "Seed treatments represent one tool on which many sustainable agriculture technologies rely upon. By protecting seeds from planting to emergence, seed treatments can improve stand establishment and increase potential yield."

Helmut Schramm, head of the seed treatment business at Bayer CropScience in Monheim, Germany, agrees, saying seed treatments deliver clear environmental, economic and social benefits, making this technology a perfect tool for sustainable agriculture.

"Innovative seed treatment technology represents an environmentally-sound approach to crop protection," he says. "Treating the seed provides a targeted and effective means of application that helps increase yields, safeguard our environment and ensure a sustainable means of crop production."

Over the years, seed treatments have evolved from simply protecting the seed to helping improve plant stand and early plant health.

"Seed treatments are increasingly designed to also enhance plant emergence, growth or nutrition efficiency, which, along with crop protection, lead to a more vigorous and uniform crop," says Schramm. "This forms the optimal base for a high-quality crop which can fully exploit its yield potential."

Greg Lamka, chair of the International Seed Federation Seed Treatment and Environment Committee, says that in order to maximize yield, it's important to have a full and uniform stand. The STEC committee was established in the 1990s to raise the seed industry's awareness about the use of different seed treatments and to promote a better understanding of how production could be improved and made more efficient. The committee consists of seed companies and crops protection companies who wish to promote the safe and effective use of seed treatment products. "Growers are paying more for the seed, so expectations are rising about the performance of our products. Seed treatments are a way to ensure that the products will perform to their maximum, based on the environment that they're put into," says Lamka, adding the vast majority of the seed used in developed countries is treated.

COMBINATION OF INGREDIENTS

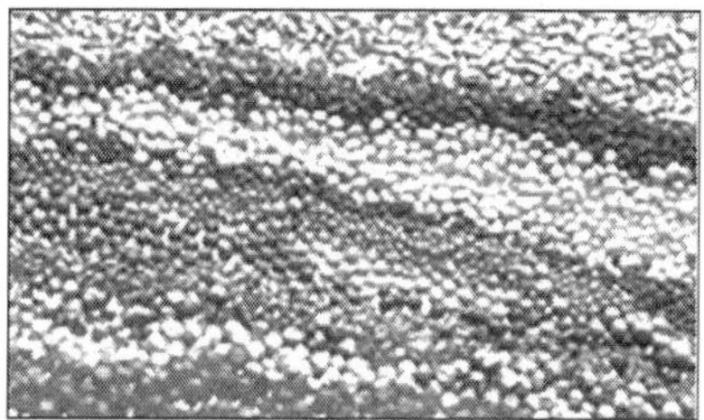

And what do seed treatment products consist of? The most important thing, of course, is the active ingredient which is responsible for the fungicidal and/or insecticidal effect. Yet a seed treatment product contains many other components too, the most important of which are adhesive substances (to make the active ingredient stick to surface of the seed), dispersion substances (to allow for even distribution of the active ingredient), and colorants (so that it can be seen immediately whether or not even distribution of the seed dressing has been achieved).

These and other important properties such as seed flowability are supported and complemented by the co-application of a proper film coating designed for a specific seed treatment product.

MODERN TECHNIQUES

Nowadays, modern seed treatment machines have been designed to a high standard for treatment of large quantities of seed. However, different seed treatment formulations and different seeds sometimes require different machines.

The most common seed treatment formulations worldwide are flowable concentrates, wettable powders and liquids. Advanced seed dressing methods include film coating and pelleting.

Bayer CropScience's most recent development is water-borne coatings, seed treatments in the form of a water-based suspension which are particularly user-friendly.

They generate neither dust nor solvent vapors during use, and machines and equipment are easy to clean with just water.

APPLICATION METHODS

SEED DRESSING

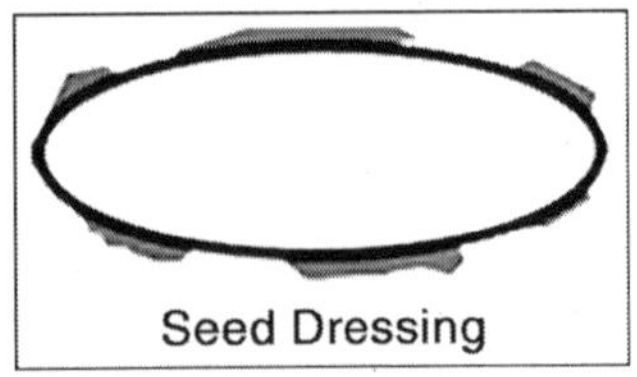

Seed Dressing

Application using simple application machinery,

FILM COATING

Film coatings are applied together with the insecticidal and/or fungicidal seed treatment product to improve the application quality. Proper film coatings applied at the correct use rate reduce abrasion, dust formation and hence the loss of active ingredient during application, packaging and sowing. They improve the even distribution of the seed treatment products on the seeds and restore good seed flow ability/sowability.

Film coatings are part of professional seed treatment for field crops and vegetables, and are applied mainly by professional seed companies. In order to confer all the technical properties such as reduced abrasion and dust formation, good flowability, and improved seed coverage and coloration, modern film coatings are quite complex and sophisticated products which contain polymers, loading materials and wetting and stability agents in combination with pigments and shine agents.

The combination of these components and the exact recipe strongly depend on the detailed requirements, the crop and the seed treatment product used. In addition, film coatings must not hinder water uptake by the seeds, their germination or field emergence. Film coatings form a very thin film and do not change the size, shape or weight of the seeds.

PELLETING

Pelleting has two main purposes. One is to give seeds with an uneven surface a uniform and homogeneous size and shape, *e.g.* for sugar beets and fodder beets.

The second purpose of pelleting is to increase the size and/or weight of very small seeds such as vegetable or grass seeds. In both cases the intention is to adapt and change the shape, size and/or weight of the seeds to allow precision sowing with modern equipment.

The inert materials used for the pelleting process must be capable of forming a robust and stable pellet but at the same time must not hinder water uptake by the seeds and hence germination and emergence. Seed treatment products are usually applied to pelleted seeds in a second step and always in combination with a film coating.

PELLETING AND COATING

The term describes the sequential application of different film coatings in combination with different seed treatment products. It can be used with either pelleted or non-pelleted seeds.

MULTILAYER COATING

A highly sophisticated method allowing sequential application of multilayer materials, including the incorporation of fungicides and insecticides.

GIVING SEEDS A FILM COATING

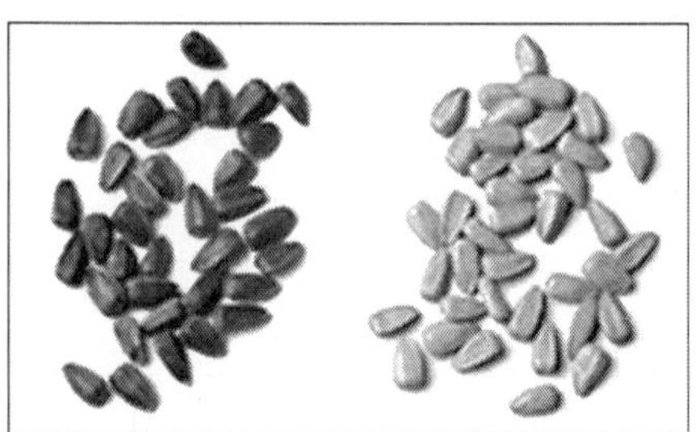

Film coating products contain polymers, dyes and surfactants and form a fine air- and water-permeable film that improves the distribution and retention of crop protection agents on the seed surface. The coating products contribute to a significant reduction of the amount of dust released during the application of the crop protection agents to the seed and during handling and use of the treated seed on the farm.

This means the operator is better protected.The film coating also improves the flowability of the seed, *i.e.* the ability of seed to flow instead of clumping in the treatment station or sower, which means easier production on an industrial scale.

Film coating also allows sowing rates to be managed accurately. The coloration of the coating gives the seed an attractive, glossy appearance, but also has the practical use of allowing differentiation between varieties and between different types of treatment. Film coatings do not significantly modify the shape and weight of the seed.

PELLETING FOR UNIFORM SIZES AND SHAPE

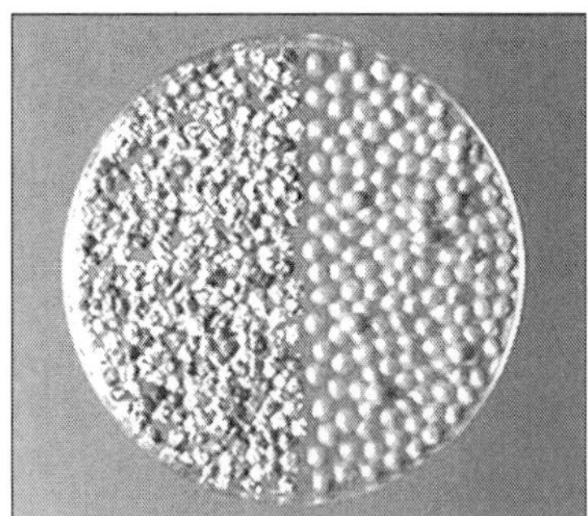

Surrounding a seed with a thick coat, the pelleting process is designed to make seeds uniform in size, shape and weight so that they can be precision-sown with modern drilling machinery. The other advantage of pelleting is the additional protection it provides to the seed. The pelleting material must be robust enough to remain in place during handling, transport and sowing. Pelleting guarantees that the active ingredients are distributed evenly among individual seeds within a seed lot. The application of a covering layer of fungicide and/or insecticide ensures that exactly the right dose is available at the right time.

ADVANTAGES

Seed treatments refer to the application of crop protection products directly to a seed to protect it from seed-borne and soil-borne pathogens and insects. The time between planting and emergence is the most vulnerable stage of plant development. Once in the soil, the seed is susceptible to damage from insects, pests, bacterial pathogens and fungi. Seed treatments allow farmers to protect against these threats during planting, says Jones.

When possible, farmers plant earlier than they used to in order to maximize yields, and often reduce tillage or decide not to till at all, says Lamka, noting that both of these practices significantly impact the seed bed.

"The earlier you plant, the more often it's going to be cold and wet. The more plant debris you have laying on the surface of the soil, the colder and wetter the seed bed will be and the more plant disease inoculums that will be present. The colder and wetter the seed bed, the slower the seed will germinate, and the slower it comes up," he says, adding that this provides more opportunity for fungi to attack and kill seed, or greatly reduce its health. This risk can be decreased by using seed treatments.

Seed treatments are also a highly targeted way of applying pesticides, notes Schramm. "Instead of spraying the entire field area, less than one per cent is treated, and so only insects and pathogens that forage on the plants are exposed," he says. "Therefore, beneficial species and other species that live on and around the plants are protected." Less product use per area also leads to decreased risk of off-crop drift, which consequently has a reduced impact on species in adjacent areas, he adds.

Lamka agrees, saying that in the past, if you had a problem with insects or disease, you would make a foliar or granular application. However, when you use seed treatment, you end up using less product and you're burying the product underground. "So we've greatly reduced our impact to the environment by using these very small amounts of focused material as a seed treatment," he says.

Many of the new chemistries used for treatments are systemic, explains Lamka. They come off the seed coat into the soil, and are absorbed by the seedling through the root system as it grows. These products often protect the seed and the seedling for approximately three weeks after emergence.

Farmers prefer seed treatment over crop spraying because it is more effective in terms of crop protection, and generates vigorous plants and increasing yields, while being more cost-efficient, says Schramm.

"Entire field spraying can be spared, reducing the use of fossil fuels (and the greenhouse gas emissions associated with their use) as some foliar sprays are no longer necessary," he says. "So this addresses the economic pillars of sustainability, while complementing the technology's environmental benefits." Seed treatments are an environmentally safe way to protect plants because of the small use of active ingredients per unit of land area, says Lamka, who is also the global senior manager of seed applied technologies for Pioneer Hi-Bred in Johnston, Iowa.

"The products are more environmentally safe than they've ever been in history. Using seed treatments is a good stewardship practice," he says.

Seed treatments go back thousands of years, notes Lamka, to when they put salt brine on wheat seed to get rid of certain seed-borne diseases. Years ago, mercury was also used as a seed treatment because it was very effective at killing insects that were attacking the seed, but it's a toxic product to all living organisms. These early toxic products have been banned and taken off the market.

"Today we're using much safer products. They are safer for the people handling them, safer for the environment and much safer for the seed itself," says Lamka.

6

Seed Germination

INTRODUCTION

Some knowledge of the biology of seeds is essential to their proper handling. The use of seed for artificial regeneration makes possible a considerable degree of control over the conditions in which it is collected, processed, stored and treated, but the seed's inherent characteristics have been evolved as a result of millenia of adaptation to natural regeneration under local conditions. Knowledge of flowering phenology enables the collector to select the timing and methods of seed harvesting most appropriate to the species, while handling, storage and pretreatment of seed will benefit from a knowledge of how seeds develop in nature.

POLLINATION AND FERTILIZATION

A seed is a reproductive unit which develops from an ovule, usually after fertilization. Ovules are borne by both the angiosperms (true flowering plants) and the gymnosperms (which include the conifers). In the angiosperms the ovules are totally enclosed within the ovary, while in the gymnosperms the ovules are "naked", typically borne in pairs on the upper surface and near the base of each scale in a female cone. Since the cone scales remain tightly closed except at the time of pollination and later at seed shed, the term "naked" is a relative one.

Seed development is initiated by fertilization, the union of a haploid male nucleus from the pollen grain with a haploid female nucleus within the ovule to form a new diploid organism. Fertilization must be preceded by pollination, the arrival of a pollen grain on the stigma of the female flower in angiosperms or close to the micropyle of the gymnosperm ovule. It is important to distinguish the two separate processes of pollination and fertilization. In most angiosperms the elongation of the pollen tube is rapid and the interval between pollination and fertilization is only a few days or even hours. In a few angiosperms (*e.g. Liquidambar,* some species of *Quercus*) and many gymnosperms (*e.g. Pseudotsuga, Larix, Picea*) the interval is several weeks or months, while in other species of *Quercus* and in many Pinus it is a year to 14 months.

ANGIOSPERM SEED DEVELOPMENT

At the time of fertilization a typical angiosperm ovule consists of one or two protective coats - the integuments - and a central tissue—the nucellus. Often the integuments and the nucellus are clearly differentiated only in the region of the micropyle - the minute pore in the integuments through which, in many species, the pollen tube enters the nucellus. The ovule is attached to the wall of the ovary by a stalk - the funicle. Meiosis of a mother cell within the nucellus, followed by several mitotic cell divisions, leads to the formation of the embryo sac, a haploid eight-nucleate, seven-celled structure which occupies the central space within the nucellus. When the pollen tube reaches the embryo sac it releases two male gametes.

One male gamete unites with one of the nuclei in the embryo sac—the egg cell - to form a zygote which later develops into the diploid embryo plant. The second male gamete unites with two other female nuclei - the polar nuclei - to form a triploid cell which later develops into the endosperm, a tissue which acts as a food reserve for the growing embryo. The remaining five nuclei of the embryo sac (2 synergids and 3 antipodal cells) play no further role in seed development. Successful fertilization of the egg cell and successful triple fusion with the polar nuclei are both necessary for development of a viable seed.

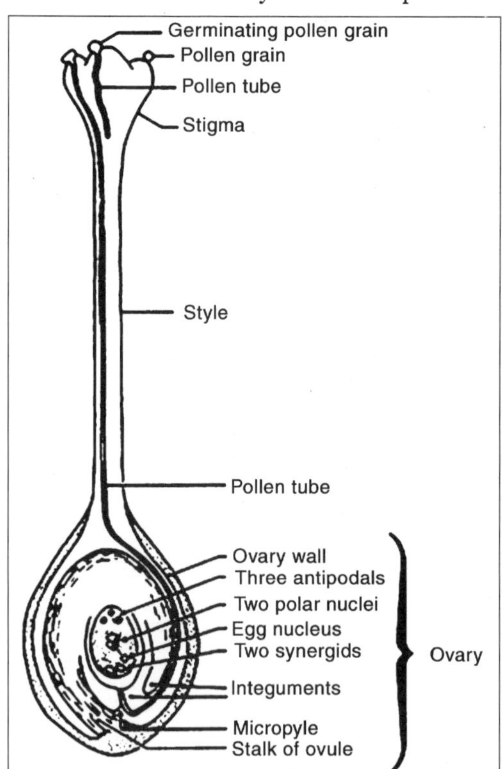

Fig. Longitudinal Section Through a Typical Pistil just before Fertilization. (USDA Forest Service)

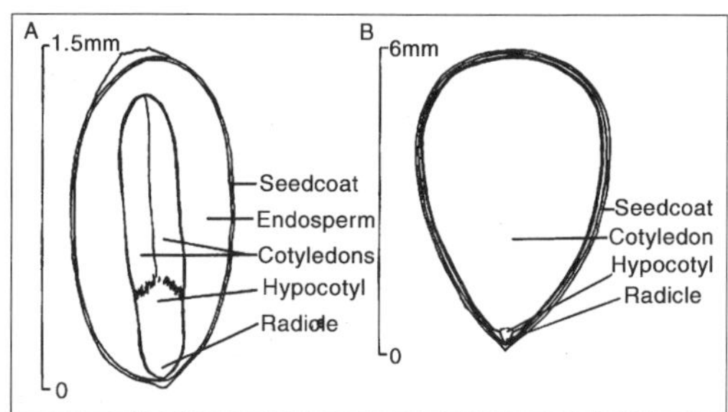

Fig. Longitudinal Sections through Ripe Seeds of:- (A) *Paulownia Tomentosa* Showing Conspicuous Endosperm. (B) *Tectona Grandis*. Endosperm has Disappeared and Cotyledons Occupy Almost the Entire Seed Cavity. (USDA Forest Service)

Examples of different types of fruits:

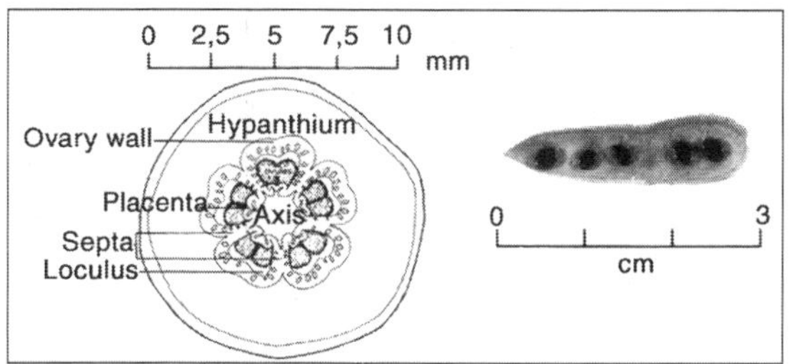

Fig. (A) Cross-section of Capsule of *Eucalyptus Preissiana* Showing Loculi, Axis, Placentae and Ovules. (Division of Forest Research, CSIRO, Australia). (B) Open pod with Seeds of *Acacia Aneura*. (FAO/Division of Forest Research, CSIRO, Australia).

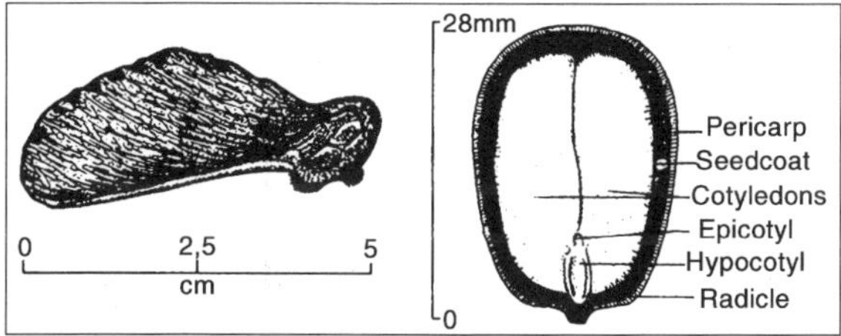

Fig. (C) Samara of *Triplochiton soleroxylon*. (Forest Research Institute of Nigeria). (D) Nut (acorn) of *Quercus rubra*. (USDA Forest Service).

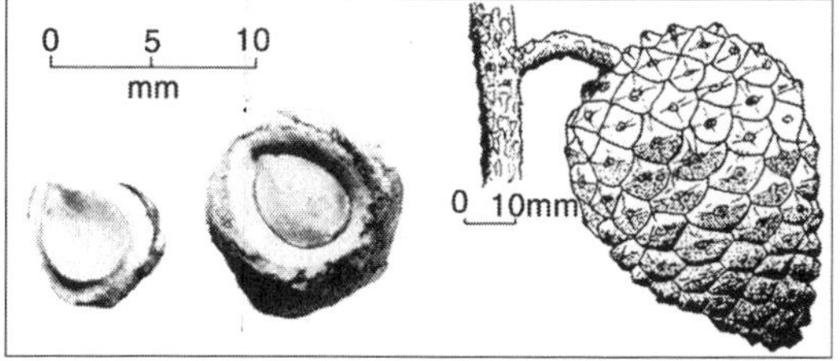

Fig. (E) Drupe of *Tectona grandis*. (S.K. Kamra). (F) Cone of *Pinus oocarpa*. (A.M.J. Robbins)

Development of the fertilized ovule into the mature seed involves several different parts.

From the outside inwards these are as follows:

1. The integuments of the ovule become the seedcoat of the mature seed. This sometimes consists of two distinct coverings, a typically firm outer seedcoat, the testa, and a generally thin, membranous inner coat, the tegmen. The testa protects the seed contents from drying out, mechanical injury, or attacks by fungi, bacteria and insects, until it is split at germination. But there is great variation in seedcoat among angiosperms.
2. The nucellus may persist in some genera as a thin layer—the perisperm -lying inside the seedcoat and supplying food reserves to the embryo. In most angiosperms, however, it soon disappears and its function is taken over by the endosperm.
3. The endosperm commonly grows more rapidly than the embryo during the period immediately after fertilization. It accumulates reserves of food and its fullest development is rich in carbohydrates, fats, proteins and growth hormones. In some species the endosperm remains conspicuous and still fills a greater part of the seed than the embryo even when the seed is ripe. In others, such asTectona, the embryo absorbs food reserves from the endosperm during its later stages of development, until the endosperm disappears by the time the seed is mature.
4. The embryo occupies the central part of the seed. Its degree of development at the time the seed is ripe varies greatly according to species. In some it is possible to distinguish all parts of the rudimentary plant - the radicle which at germination will give rise to the primary root, the seed leaves or cotyledons, the plumule from which will develop the primary shoot, and the hypocotyl which connects the cotyledons with the radicle. If the embryo absorbs all the food reserves from the endosperm, the thick fleshy cotyledons commonly become the main organs for food storage and occupy almost the whole of the seed cavity.

Although the storage function within the embryo is normally performed by the cotyledons, in *Anisophyllea, Barringtonia* and *Garcinia* it is completely taken over by the swollen hypocotyl which fills the seed cavity; the cotyledons are vestigial or absent. This is also the case in *Lecythis* and *Bertholletia*, hence the edible content of a Brazil nut is neither endosperm nor cotyledon but hypocotyl.

In some species the embryo is still small and undeveloped when the seed is ready for dispersal, and an additional period under suitable environmental conditions is needed for maturation of the embryo after seed shed, before the seed can become capable of germination, *e.g. Fraxinus excelsior*. At its most complex the ripe seed may thus consist of diploid tissue from the mother tree (the seedcoat, including testa and tegmen, and the perisperm), triploid tissue in the endosperm, and diploid tissue of the new genetic combination in the embryo offspring. But both perisperm (nearly always) and endosperm (not infrequently) may be missing. The essential constituents of all seeds are the embryo, the protective covering of the seedcoat and a reserve of food substances which may be stored in the cotyledons, hypocotyl, endosperm or perisperm.

Occasionally more than one embryo may develop in a single seed and such polyembryony has been reported from several genera. It is, however, the exception.

ANGIOSPERM FRUIT DEVELOPMENT

Development of the fertilized seed is normally accompanied by development of the fruit. In the simplest case the ovary wall becomes thickened to form the pericarp.

This may be:

- Dehiscent, splitting open when ripe to release the enclosed seeds; examples are the *capsule* (*e.g. Eucalyptus*), a multilocular fruit derived from a syncarpous ovary, and the leguminous pod (*e.g. Cassia*), which is derived from a single carpel and spilts along two sutures. The pericarp may be dry, semi-fleshy or fleshy at the time of dehiscence. Semi-fleshy to fleshy capsules are common in the humid tropics (*e.g. Baccaurea, Durio, Dysoxylum, Myristica*) and are often associated with the development of variously coloured, tasty or smelly pulp (aril or sarcotesta) around the seed.
- Indehiscent or dry, closely fused with the seed; examples are the *achene,* a small hard one-seeded fruit with membranous pericarp, the *samara,* similar to the achene but with pericarp extended to form a wing (*e.g. Triplochiton)* and the *nut*, a rather large one-seeded fruit with woody or leathery pericarp (*e.g. Shorea, Quercus).*
- Indehiscent and fleshy, often distinguished by colour, smell and taste to attract fruit-eating birds and animals. Two types are distinguished. The *berry* has an outer skin and inner fleshy mass, containing seeds that have a hardened seedcoat (*e.g. Diospyros,Pouteria*). The *drupe* has the inner layer of the pericarp hardened to protect the seeds (*e.g.* P*runus, Gmelina, Azadirachta, Mangifera)*; the seedcoat, having no protective function in a drupe, is usually papery or membraneous. The different pericarp layers in a typical drupe are known as exocarp (the skin), mesocarp (the flesh) and endocarp (the stone). The stone may be actually stony as in *Gmelina* or leathery as in *Mangifera*.

In some species other parts of the flower, as well as the ovary wall, take part in fruit formation. An example is the pome, found in apples and pears, in which the enlarged fleshy receptacle forms the greater part of the fruit, while the pericarp forms the core. An additional partial or entire protective covering may be provided by fused bracts arising below the flower - the involucre. This may be papery, as in *Tectona*, or thicker and leathery as in the "acorn cup" of *Quercus*. Some fruits are formed by the coalescing of an entire inflorescence *e.g. Morus, Chlorophora,Anthocephalus, Artocarpus.* At the opposite extreme, in several genera of the *Sterculiaceae* (*e.g. Fimiana, Pterocymbium and Scaphium)*, fruit formation does not occur at all in the normal angiospermous manner. Soon after fertilization, the carpel (follicle) splits on one side and develops into a large membraneous scale-like or boat-shaped wing; the fertilized ovule develops in a naked position at or near the base of the open carpel, in a gymnospermous manner. Such fruits must be the most primitive of all angiosperm fruits. At maturity the seeds are dispersed, attached to their carpels which now behave as wings.

The interval between flowering and maturation of seeds and fruits varies greatly with species, even within the same genus. In *Eucalyptus* it varies from one month in *E. brachyandra* to 10–16 months in *E. diversicolor*. In most Malaysian Dipterocarps it is between two and five months. In *Tectona grandis* it takes 50 days from flowering for the green fruits to develop to full size but 120–200 days before they are fully ripe.

In a study of rooted cuttings of *Gmelina* arborea in pots in Nigeria, individual flowers took 11 days from flower bud to opening and 45 days from flower bud to ripe fruits. In *Pterocarpus angolensis* the interval between flowering and fruit maturation is 8 months. The shortest interval on record between flowering and seed maturation for a tropical timber species is apparently 3 weeks, for *Pterocymbium javanicum*. In contrast, some species of temperate *Quercus* take about 18 months from flowering to production of mature seeds.

In most species fertilization of one or more ovules must precede fruit formation. In a few species, however, fruits are set and mature without seed development and without fertilization of an egg. Such fruits, called parthenocarpic fruits, occur in several genera of forest trees including *Acer, Ulmus, Fraxinus, Betula, Diospyros and Liriodendron* (Kozlowski 1971). Mature fruits do not invariably indicate mature seed, still less can the number of sound seeds be predicted from the number of fruits. In Tectona the number of sound seeds per fruit can vary between 0 and 4 and still greater variation is possible in other genera.

SEED DISPERSAL IN ANGIOSPERMS

There is thus an immense variety among angiosperm fruits. Much of it is related to the need for seed dispersal. Survival and growth of young seedlings under the parent tree is often difficult, because of lack of light and intense root competition. Dispersal over a wide area can ensure that some seeds find conditions suitable for germination and survival, even though the vast majority will perish from the effects of harsh site conditions, competition or destruction by animals or disease.

Dispersal by wind is assisted when the seeds are very light and small *e.g. Eucalyptus,* or when either the seedcoat *(Salix, Ceiba, Dyera)* or the pericarp *(Triplochiton, Pterocarpus, Koompassia, Casuarina, Fraxinus)* possesses wings or hairs which serve to prolong flight. Fruits may also be winged by the enlargement of persistent sepals (most Dipterocarps) or persistent petals.

The distance of seed or fruit dispersal by wind depends not only on the weight and type of dispersal unit but also on the local wind conditions and the exposure and isolation of the mother trees. Studies of the winged fruits of *Shorea contorta* in the Philippines indicated that 90 per cent of the fruits travelled 20 m or less from the stem of the mother tree (Tamari and Jacalne and a summary of other dipterocarp studies compiled by the same authors shows that most fruits landed within 30 m or, at most, 40 m. This compares with a dispersal distance within 2–3 m of the crown perimeter for a heavy, wingless seed such as *Quercus crispula* in Japan and a distance of over 60–90 m for 5 per cent of the light, winged seeds of *Betula ermannii* downwind from a belt of mother trees left in a logged over area.

Fleshy edible fruits and arillate seeds, on the other hand, encourage dispersal by birds or mammals. When such fruits or seeds are eaten by animals, the seeds, protected by the hard seedcoat or endocarp, often pass unharmed through the digestive tract and are deposited in the faeces at a considerable distance from the place where they were consumed. In many cases the digestive juices actually assist subsequent germination through softening of the hard seedcoat. In Africa the hornbill is a highly efficient dispersal agent for seeds of *Maesopsis eminii.* Sometimes the process is so effective as to be an embarrassment. Free-ranging goats eat the pods of *Prosopis* in some countries and spread the seeds indiscriminately over large areas; excellent germination and the aggressive pioneering qualities of the young plants may then render this genus a dangerous weed-tree. Coralling of the goats and collection of the seeds for use under strictly controlled management can overcome the problem. In other cases the fruit is eaten while the stones or seeds are rejected, but the animal may carry the fruit some distance from the parent before dropping the seeds. Rodents remove and store nuts or seeds; many are subsequently eaten but some may escape to germinate in the new situation.

Wind and animals are the most important agents of dispersal, but dispersal by water is common in some riverine species and large and heavy fruits are distributed to some extent by gravity on steep slopes.

GYMNOSPERM SEED DEVELOPMENT

Gymnosperm ovules have certain characteristics in common with angiosperm ovules, but there are a number of differences. There is normally a single protective integument which in a typical female cone is partially fused to the ovuliferous scale carrying the paired ovules. Within the integument is the nucellus which at fertilization, as in angiosperms, is clearly separated from the integument only in the region of the micropyle. Meiosis within the nucellus, followed by mitotic cell divisions, leads to the formation of a multicellular haploid tissue—the female gametophyte. By the time of fertilization it has developed much further than the 8-nucleate embryo sac in the angiosperms and has largely displaced the nucellus. At its micropylar end it is differentiated into one to many archegonia, each of which contains a large egg cell.

At fertilization the pollen tube releases two male nuclei into an archegonium, one of which unites with the egg nucleus. The resulting zygote later develops into the new diploid embryo. The second male nucleus aborts in *Pinus* but may fertilize a second archegonium in other genera *e.g. Cupressus*. It never unites with female polar nuclei to form a triploid tissue analogous to the endosperm of angiosperms; this type of tissue is unknown in gymnosperm seeds. For detailed descriptions of gymnosperm embryogeny, readers are referred to the specialized literature.

The mature seed consists of some or all of the following:

- The seedcoat or testa developed from the integument, diploid from the female parent.
- The diploid perisperm, developed from the nucellus. In most species this is absorbed by the female gametophyte and has disappeared by the time the seed is ripe, but it is still recognisable as a distinct tissue in *e.g. Pinus pinea*.
- The haploid female gametophytic tissue which serves as a food storage organ to nourish the embryo. Its function is the same as the endosperm in angiosperms and it is frequently called by that name, though this usage has been deprecated.
- The embryo, with the same parts of radicle, cotyledons, plumule and hypocotyl as in angiosperms. The number of cotyledons varies between and within genera, being up to 18 in *Pinus,* compared with the constant two in the dicotyledons which comprise the great majority of angiosperm trees. The essential constituents of embryo, protective covering and food storage tissue are present in all gymnosperm, as in all angiosperm, seeds.

More than one archegonium may be fertilized within a single ovule, but in the great majority of cases only one embryo per seed develops to maturity. Polyembryony does occur but is uncommon in most genera.

GYMNOSPERM FRUIT DEVELOPMENT

After fertilization the female cone which is typical of several important gymnosperm genera *e.g. Pinus, Picea, Pseudotsuga, Araucariain* creases in size and weight, in moisture content and accumulated food reserves. As the cones approach maturity, the moisture content decreases again, accumulated food reserves move from cone to seed and the cone becomes more or less woody.

In *Pinus* a thin membranous flake becomes detached from the ovuliferous scale and adheres to the ripe seed, forming a wing. In *Juniperus* the cone scales grow together to form a fleshy berry-like fruit, while in *Podocarpus* and *Taxus* each singly borne seed becomes partly enclosed in a brightly coloured cup, the aril. The woody cone is, however, the most characteristic type of fruit in gymnosperms.

As in angiosperms, there is wide variation in the interval between flowering and seed maturity and dispersal. Because of the lengthy gap between pollination and fertilization in pines, mentioned earlier in this chapter, the total period between pollination and cone maturity is usually about two years in this genus; among tropical pines, average periods are 23 months in *Pinus kesiya,* 18–21 months in *P. oocarpa*. In *Agathis robusta* it takes 16 months from pollination to cone maturity, in *Araucaria cunninghamii* up to 24 months, in *Araucaria hunsteinii* 21–24 months. In several temperate genera development is completed within a single season *e.g.* in 5 months in *Pseudotsuga menziesi.*

SEED DISPERSAL IN GYMNOSPERMS

In the typical gymnosperm cone, ripening and drying of cone and seed causes the cone scales to open and release the seeds. Dispersal is by wind, assisted by the presence of seed wings in some genera *e.g. Pinus.* In some species of pine, the "closed-cone pines" *e.g. P. radiata*, there is usually an interval of months or years between ripening of cone and seed and the opening of the cone to release the seeds. In a few cases, such as the interior provenances of *Pinus contorta,* the cones open only when subjected to the heat of occasional fierce forest fires. On the other hand the cones of *Abies* and *Araucaria* disintegrate readily on the tree within a few weeks of ripening.

Seed dispersal by animals is less common, but the "berries" of *Juniperus* and the fleshy fruits of *Podocarpus* are examples. In addition seeds of temperate conifers are collected and stored by rodents and some may germinate before being eaten.

PROCESS

At one extreme, certain species of mangrove are viviparous, the seeds germinating before they separate from the parent. At the other extreme, seed of some species may remain dormant but alive for many years, capable of germinating if an event occurs to break the dormancy state. The subject of dormancy is discussed later in this chapter. Between the viviparous and the deeply dormant seed types occur many types of seed which are capable of germination soon after seed shed provided that environmental conditions are suitable. Just as fertilization initiates the transformation of the ovule into the ripe seed, so does germination transform the embryo within the seed into the independent seedling. For the purpose of laboratory testing, germination is defined as the emergence and development from the seed embryo of those essential structures which are indicative of the seed's capacity to produce a normal plant under favourable conditions.

At maturity and seed shed many seeds have lost the greater part of the moisture which they contained in earlier stages. For example the embryo and female gametophyte of *Pinus lambertiana* contain as much as 50 per cent moisture content (fresh weight basis) shortly after fertilization, but by the time of natural seed dispersal the moisture content of the embryo is reduced to 23 per cent and of the female gametophyte to 38 per cent. Reduced metabolic activity is associated with the drying of the seed, so that the embryo is in a temporarily resting or quiescent state, which in non-dormant seeds can be easily reactivated by suitable conditions.

These conditions are:

- Adequate moisture
- Favourable temperatures
- Adequate gas exchange and, for some species,
- Light.

There is considerable variation between species in the optimum levels of the different factors and there is frequently an interaction between them.

Germination consists of three overlapping processes:

1. Absorption of water mainly by imbibition, causing a swelling of the seed and eventual splitting of the seedcoat,
2. Enzymatic activity and increased respiration and assimilation rates which signal the use of stored food and translocation to growing regions,
3. Cell enlargement and divisions resulting in emergence of radicle and plumule.

In most seeds the radicle of the embryo is close to the micropyle, where absorption of water is easier and quicker than through the seedcoat. As the radicle swells, it exerts pressure on the seedcoat which commonly splits first at this point to free the radicle. This gives rise to the primary root which grows down into the soil and soon produces lateral roots. Subsequent stages depend on whether the species exhibits epigeal germination *e.g. Pinus*—the hypocotyl elongates and the cotyledons are lifted above ground - or hypogeal germination *e.g. Quercus* —the hypocotyl is undeveloped and the cotyledons remain on or in the ground. In hypogeal germination the cotyledons can have only a food storage function, or a haustorial function (in species in which food is stored in the endosperm *e.g.* palms, *Scorodocarpus)*, while in epigeal germination they may also perform a valuable photosynthetic function during early growth of the seedling.

In epigeal germination anchoring of the young plant by the radicle is followed by rapid elongation of the hypocotyl which arches upwards above the soil surface and then straightens; simultaneously the cotyledons and plumule are exposed, to which the seedcoat may or may not still be attached. The plumule then develops into the primary shoot and photosynthetic leaves. In the subtype "durian germination" the hypocotyl elongates but the cotyledons are shed while still enclosed within the seedcoat. In hypogeal germination, the cotyledons remain *in situ* underground or on the ground while elongation takes place in the plumule. In the subtype "semihypogeal germination" the cotyledons are exposed but remain on the ground. The two main types, epigeal and hypogeal and the two subtypes, durian and semihypogeal are the result of the four possible combinations of two independent variables: hypocotyl elongated or not and cotyledons exposed or not. All four combinations occur in the humid tropics.

Even in non-dormant seeds there is considerable variation, between species and individuals, in the speed of germination, from a few days to several weeks; much of this is due to different rates of imbibition in the first stage. Many tropical rain forest species, with high moisture content and permeable seedcoats at seed fall, must germinate within a few weeks. If they fail to find suitable conditions soon, they lose viability and die.

DORMANCY

The term "dormancy" refers to a condition in a viable seed which prevents it from germinating when supplied with the factors normally considered adequate for germination - suitable temperature, moisture and gaseous environment. A viable seed is defined as one which can germinate under favourable conditions, providing any dormancy that may be present is removed.

Dormancy in nature serves to protect the seed from conditions which are temporarily suitable for germination but which quickly revert to conditions too harsh for survival of the tender young seedling. Thus a seedcoat relatively impermeable to moisture prevents germination during isolated showers in the middle of a long dry season, while permitting it during a sustained rainy season. In the cool temperate zone the type of embryo dormancy which can be removed only through exposure to low temperatures facilitates subsequent germination in spring, while preventing it in autumn, when the resulting seedling would be unlikely to survive the winter. The strength of dormancy has been observed to vary according to latitude and provenance, and from year to year even in seed from the same parent. There is also differential dormancy within the same species and seedlot, so that germination is staggered over a more or less extended period of time. In the Malaysian woody flora, about 50 per cent of species complete germination within six weeks or less, so that the differential is not very great, but in some species such as the hard-seeded leguminous *Parkia javanica* the germination period may extend from 1 week after sowing for the first seed to two years for the last. Differential dormancy and staggered germination insures against the entire seed crop being destroyed by a single climatic catastrophe or pest.

In nature a number of external factors may work, more or less slowly, to end seedcoat dormancy. These include alternate heating and cooling, alternate wetting and drying, fire and the activities of animals, soil organisms, fungi, termites and other insects. Dormancy due to embryo immaturity will come to an end if the embryo is given time and suitable conditions in which to mature after seed shed.

The exact mechanisms of physiological dormancy of the embryo, and of the processes which can terminate it, have been widely investigated but underlying causes are still little understood. There is good evidence that growth-promoting hormones, of which gibberellin is a well known example, and growth-inhibiting hormones interact in the maintenance or breaking of dormancy. In temperate climates the balance between inhibitors and growth-promoters is altered by a combination of low temperature and high moisture maintained over a period of time which varies from species to species. This combination is provided naturally during winter, the season least suitable for growth. It can trigger biochemical changes in the embryo which lead to the breaking of dormancy, the initiation of embryo metabolism and growth, and the subsequent germination of the seedling.

Research on the physiology of tropical tree species has, unfortunately, been only a fraction of that done on temperate species. There is no reason to suppose that the combination of low temperature and high moisture—the technique of "stratification" when applied artificially—would have any effect on tropical seeds possessing embryo dormancy (if they exist). In the dry tropics, on the contrary, a combination of conditions typical of the season least favourable for growth - high temperature and low moisture—would seem more likely to trigger the breaking of embryo dormancy and lead to germination in the following rainy season. In fact, seedcoat dormancy by itself appears to provide adequate protection for species of the dry tropics.

From the forester's point of view dormancy has some disadvantages. Delayed and irregular germination in the nursery is a serious constraint on efficient nursery management. Much research has therefore gone into devising effective artificial treatments to remove dormancy, in order to ensure that the seeds germinate quickly and evenly in the nursery beds.

On the other hand, dormancy confers certain advantages. Not only does it improve the chances of survival in nature, as mentioned previously, but it preserves the seed against temporarily unsuitable conditions such as may occur during the period between seed collection and storage. High quality orthodox but non-dormant seeds, dried to the appropriate moisture content and stored at the correct temperature, should have as long a life in storage as dormant seeds, but dormancy provides an insurance against the loss of viability during transport and processing which can easily occur in non-dormant seeds in less than ideal conditions.

HAZARDS OF SEED PRODUCTION

Both the quantity and the quality of seed crops may be greatly affected by external factors. Climatic factors can affect the abundance of flowering and thus indirectly of seed production. There is some evidence that above-average temperatures and a modest degree of moisture stress in spring and early summer can induce abundant formation of flower buds in temperate regions. In Nigeria good seed years occur in *Triplochiton scleroxylon* following a particularly dry (30 per cent or less of average rainfall) August, the month when there is a diminution of rainfall between the heavy early and heavy late rains.

More extreme cases of unseasonal climate usually reduce the crop of flowers or fruits. Late spring frosts in temperate regions kill flowers or young fruits and abnormally high temperatures or drought may have a similar effect. Even if death and premature shed of whole fruits does not occur, a proportion of the seeds may abort later. Mechanical destruction of flowers or fruits may be caused by exceptionally high winds or hailstorms. Continuous rain at the time of pollen dispersal has a particularly adverse effect on the amount of seed set, whether pollination is by wind or insects. *Tectona* flowers during the rainy season and this may account for the low average rate of fertilization of 1–3 per cent reported from Thailand over the period 1967–72. Rain discourages flight of the pollinating insects as well as washing off pollen grains from the stigma before they germinate. Continuous wet weather during the pollen dispersal season is considered the main factor responsible for the characteristically poor seed crops of *Pinus merkusii* in Indonesia and Malaysia. Tamari reported that over 90 per cent of Dipterocarp flowers in Malaysia failed to develop into fruits.

Birds, mammals, insects, fungi and bacteria all do damage in flowering as well as fruiting stages. Insects are probably responsible for the most serious losses in the greatest number of species. For example, the weevil *Apion ghanaense* destroys a large part of the flowers and seeds of *Triplochiton* each year. The larvae of *Pagyda salvaris* may destroy as much as 90 per cent of the flower buds of *Tectona* in some years. Two species of the Bruchid genus *Amblycerus* can destroy many seeds of *Cordia alliodora*, but damage can be reduced by collecting the seeds three weeks before natural seedfall. The weevil Nanophyes sp. may attack up to 60 per cent of the seeds of *Terminalia ivorensis.*

Cone worms of the genus Dioryctria have been known to damage 60 per cent of maturing cones and seeds of Pinus elliottii and P. palustris in the southern USA and the same genus can also cause severe damage on *Pinus merkusii* seeds in the Philippines. Larvae of *Agathiphaga*, a genus of moths, may destroy over 50 per cent of the seeds in cones of several species of *Agathis* in Queensland and the western Pacific islands. Seeds of many dry area species of *Acacia* and *Prosopis* suffer serious damage from Bruchid larvae. Birds and mammals, especially squirrels, may consume considerable numbers of seeds in some years, although they also perform a useful service in seed dispersal. Losses from pests and diseases do not normally have a serious effect in years of abundant seed production, but in years when flowering is poor for climatic reasons they can convert a light crop into complete failure.

7

Seed Drying

INTRODUCTION

This is the reduction of seed moisture content to the recommended levels for seed storage, using techniques which will not be detrimental to seed viability. Drying of seed lots, *i.e.,* lowering down the seed moisture content to safe moisture limits, is very important in order to maintain seed viability and vigour, which may otherwise deteriorate fast due to mould growth, heating and increased micro-organism activity.

The other advantages of seed drying are:

- Permits early harvest;
- Permits long-term storage;
- Permits more efficient use of land and man power;
- Permits use of plant stalks as green fodder; and
- Permits seeds men to sell a better quality product

REQUIREMENT OF SEEDS TO BE DRIEDUP

Seeds which are dry will retain their viability for longer periods of storage in genebanks. It is recommended that, in general, seeds should be dried to between 3-7 per cent moisture content for long-term storage, except in certain cases where it has been shown that low moisture content causes problems *e.g.* soyabean should be dried to about 8 per cent moisture content. Drying has an independent effect from temperature on viability during storage and adequate drying could prolong viability for reasonably long periods without cold storage.

WHEN SHOULD SEEDS BE DRIED

The drying process should begin as soon as possible after the receipt of the seeds to avoid unnecessary deterioration. It can take up to several weeks to lower the moisture content to the low levels required for good storage.

PROCESS

Several methods are available for drying seeds. Some are more suitable in certain environments and safer for the viability of the seeds than others.

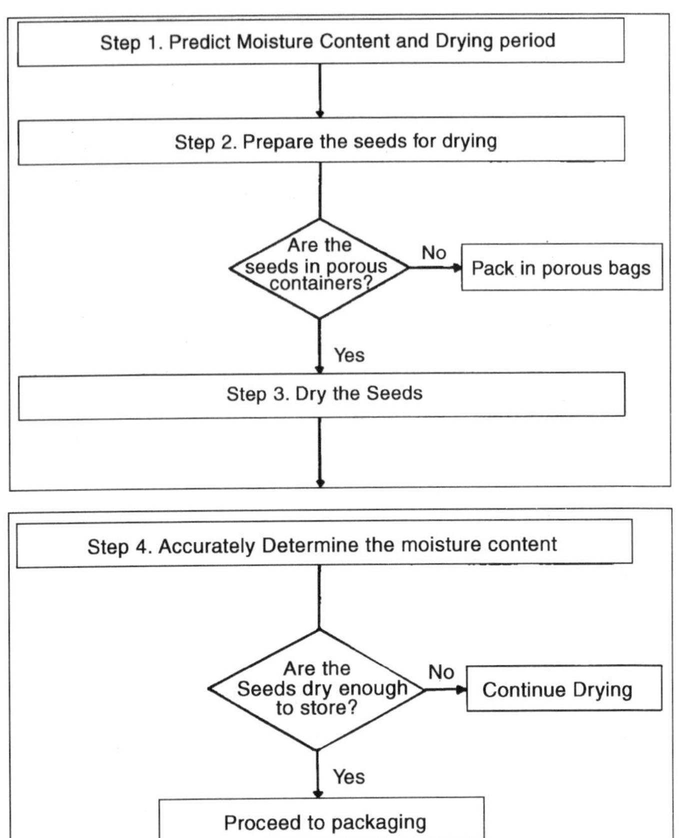

Drying seeds in an atmosphere of reduced relative humidity is recommended. The lower the humidity the faster the seeds will dry and the lower their final moisture content. A relative humidity of 10 - 15 per cent and a temperature of 15 °C are recommended by the IBPGR Advisory Committee on Seed Storage: try to obtain these conditions as closely as possible.

PREDICTION OF DRYING PERIOD

If there is no previous experience of drying seeds of a particular species it may be necessary for you to do some experimental work to predict the approximate drying period. Seeds dry at an exponential rate until the equilibrium moisture content is reached. The rate of drying of different seed lots of the same species is fairly constant for seeds dried under the same environmental conditions. The length of the drying period can be predicted in one of two ways:

PREDICTION OF THE CORRECT DRYING PERIOD BY WEIGHT LOSS

1. Predict the current percentage moisture content and the percentage moisture content required for storage.

2. Weigh the seed sample.
3. Use these three values to calculate the weight of seeds at the required moisture content by using the following formula:

$$\text{Final seed weight} = \text{Initial seed weight} \times \frac{(100 - \text{Intial\% moisture content})}{(100 - \text{Final \% moisture content})}$$

4. Weigh the seed sample at regular intervals during the drying period until the weight of the seeds has reached this calculated value.

NOTES AND EXAMPLES

The prediction of the drying period will be more accurate if the initial moisture content is determined. However, this is a waste of seeds when moisture content can be adequately predicted. 1000 g of seeds with 12 per cent moisture content were dried to 5 per cent moisture content. What would be the weight of these seeds after drying? Substitute the values in the equation on the facing page:

$$\text{Final seed weight} = 1000 \times \frac{(100 - 12)}{100 - 5} \text{g} = 926.3\text{g}$$

Therefore when the initial 1000 g of seeds have been dried to 926.3 g, their moisture content will have decreased from 12 per cent to 5 per cent.

EQUIPMENT

- Coarse balance

PREDICTION OF DRYING PERIOD FROM MEAN DRYING CURVES

1. Do not waste valuable seeds for this, but use either excess seeds or those which are being discarded because they have lost viability.
2. Take at least two lots of seeds of the species and place to dry, using the method that you will use in practice.
3. Remove a sample of seeds and do an accurate determination of moisture content for each seed lot as described in Section III.
4. The mean of the two tests can be used as a guide because other seed lots of the same species should dry at a similar rate.
5. Repeat the determination daily and plot a graph of the drying curve (mean percentage moisture content against time) for that species under these drying conditions.
6. The work can be repeated with seeds of all species which are of interest and their drying curves plotted for different conditions.

NOTES AND EXAMPLES

If the seeds are very small and are dried quickly, the moisture content can be determined more frequently during the drying period to plot a more accurate graph.

EQUIPMENT

- Grinder
- Heat resistant dishes with covers
- Analytical balance
- Forced draught oven
- Desiccator
- Silica gel
- Tongs and oven cloth

USING DRYING CURVES

1. Use the graph that you have prepared for seeds of a particular species being dried under those conditions.
2. Predict the current percentage moisture content of the accession by using the table of equilibrium moisture contents. Select the final percentage moisture content that is required for storage of this species.
3. Find the value for each of the percentage moisture contents on the vertical 'Y' axis, follow this across to the curve and read off the day on the horizontal 'X' axis for each of the moisture contents. The difference between these two values is the drying period in days.

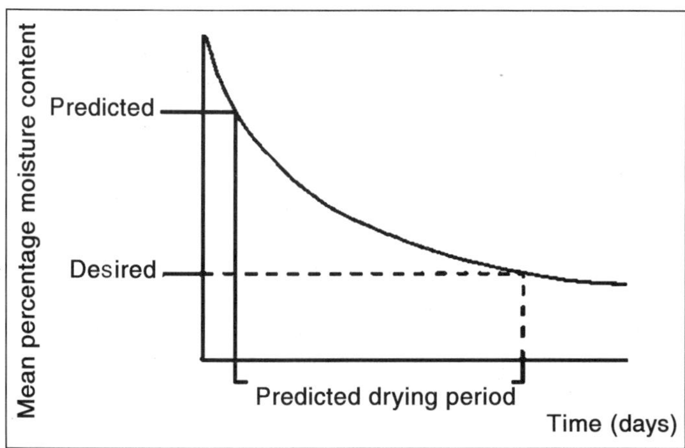

Fig. Typical Drying Curve

Notes and Examples

Example of how to predict the drying period by using a typical drying curve for small seeds of crops such as onion or cabbage, dried quickly under low relative humidity conditions:

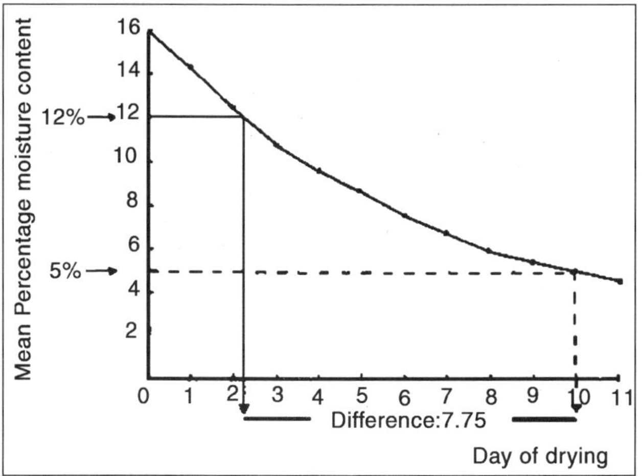

Fig. Typical Drying Curve of Small Seeds

Predict the moisture content by using the table of equilibrium moisture contents for your genebank which you have prepared. As an example take this as 12 per cent and the desired moisture content after drying as 5 per cent. Using the graph, the intersects from the curve to the time axis show the values 10 and 2.25 days.

Therefore, the time to dry the seeds to 5 per cent moisture content under the same conditions will be the difference between 10 and 2.25 days. This can be approximated to the nearest day, in this case 8 days.

In practice it is unlikely that using the graph will give you a whole number of days, therefore you should approximate to the nearest day.

Graphs of mean Drying Curves in your Genebank

Plot graphs in the space below to use for future reference:

1. Species:
   Temperature:
   Relative humidity:

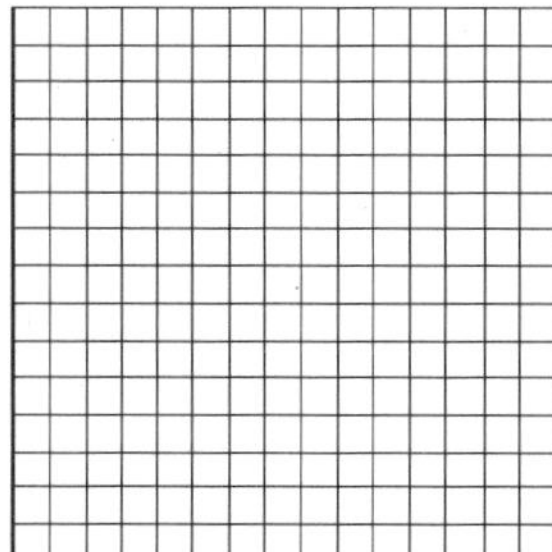

2. Species:
   Temperature:

Relative humidity:

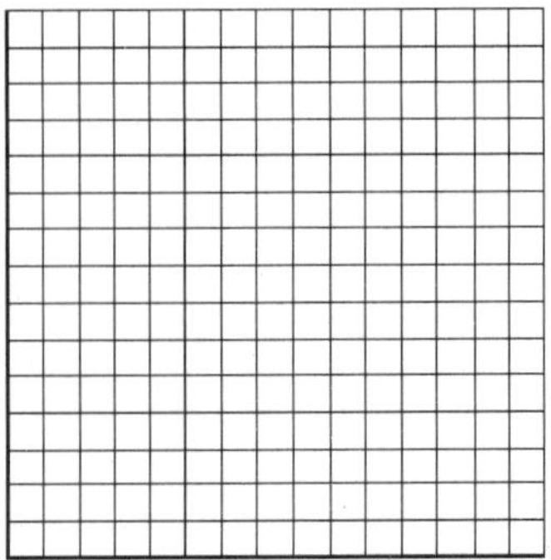

3. Species:
   Temperature:
   Relative humidity:

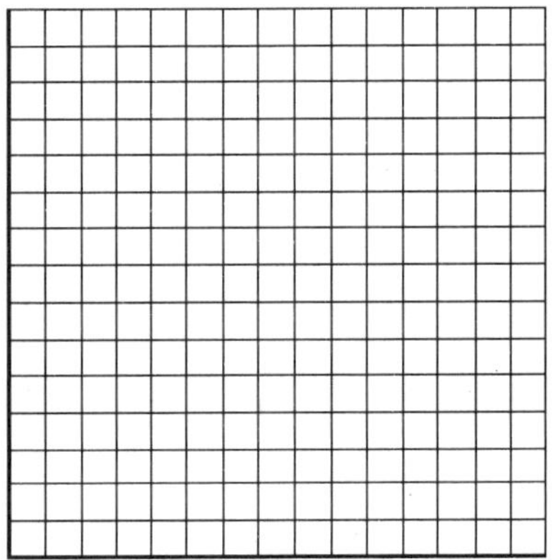

4. Species:
   Temperature:
   Relative humidity:

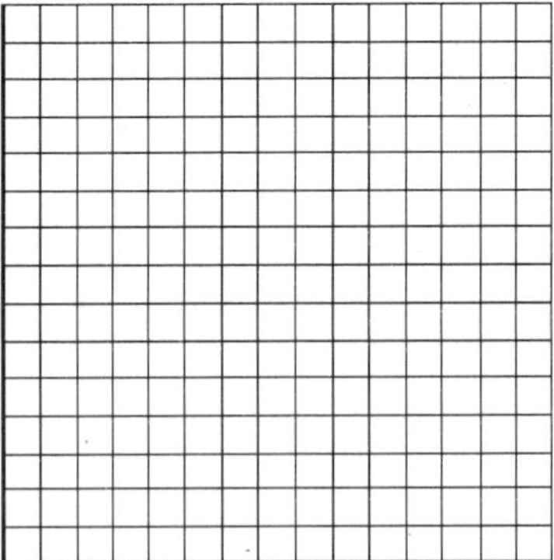

5. Species:
   Temperature:
   Relative humidity:

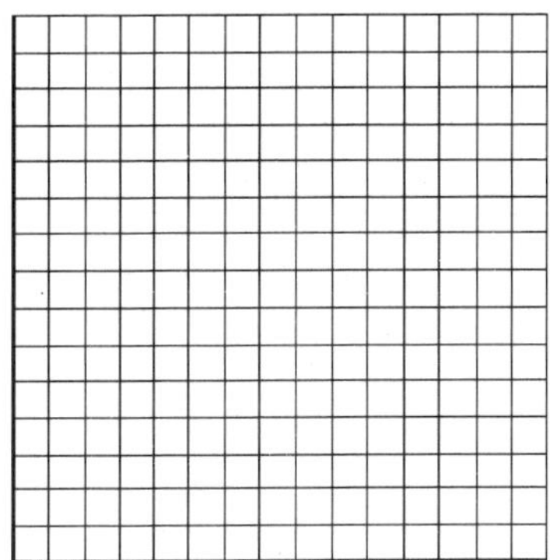

6. Species:
   Temperature:
   Relative humidity:

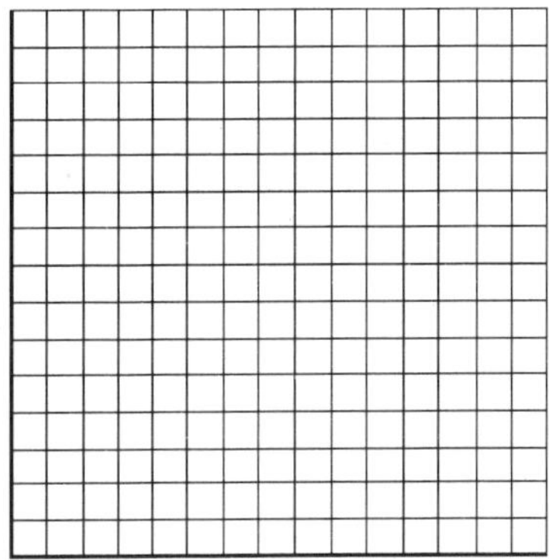

7. Species:
   Temperature:
   Relative humidity:

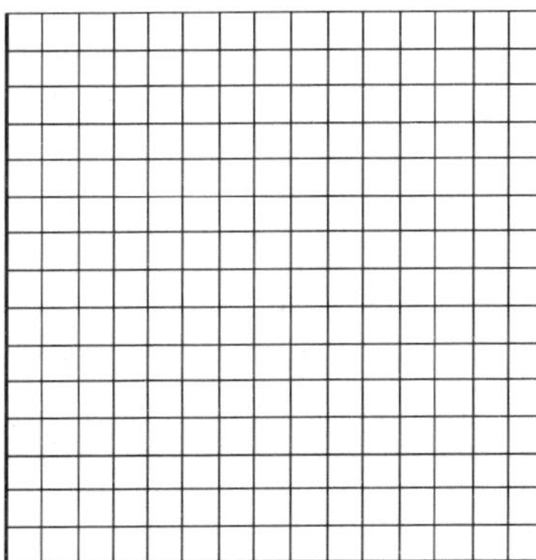

8. Species:
   Temperature:
   Relative humidity:

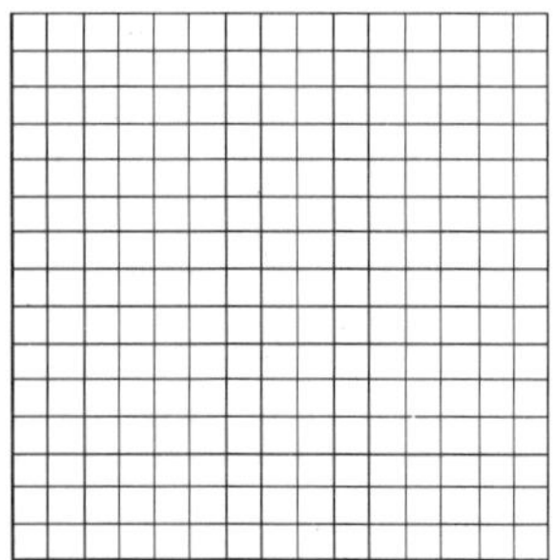

PROCEDURES

PREDICT MOISTURE CONTENT AND DRYING PERIOD

1. Use the experience and techniques already described in Section III to predict the moisture content of the accessions.
2. Predict the length of drying period required from your previous experience of drying seeds of the same species with similar moisture contents, as already described. Information on drying periods can also be found in Cromarty, Ellis and Roberts.

Notes and Examples

Predict the moisture content of each accession and the number of days to reach the required moisture content. Remember that seeds will dry more slowly as the moisture content decreases towards the equilibrium moisture content. At low moisture contents a difference of one or two days of drying will not have a large effect on seed moisture content.

PREPARE THE SEEDS FOR DRYING

1. Place each accession in labelled paper envelopes or cloth bags.
2. Do not put too many seeds in the same bag. One accession can be split into several labelled bags to help rapid drying.

Notes and Examples

Make sure that the bags are porous. Seeds will not dry in plastic bags!

DRY THE SEEDS

Several methods are available for drying seeds. The safest methods rely on leaving the seeds in an environment of low relative humidity and allowing the seed moisture content to reach equilibrium with this at relatively low temperatures. Seeds will equilibrate with the relative humidity at different rates, depending on species, seed size and conditions. Seeds dry quicker when first placed in the room and more slowly as the low moisture contents suitable for storage are approached. Two common and safe methods are described here:

Notes and Examples

Choose the best drying method for your own genebank. In tropical countries with high relative humidities, it will be more difficult to hold a drying room at very low relative humidity. A combination of methods is possible. Silica gel can be used after the first drying period if the relative humidity is too high to allow the seeds to equilibrate to low moisture contents.

Do not use sun drying because it is believed to affect long term seed viability of some species. Heated air drying should also be avoided for the same reason.

Aim for 10 - 15 per cent relative humidity and 15 °C as the optimum drying conditions.

METHODS

Drying seeds to the desired moisture level for storage may take a few days to several weeks, depending upon the species, the atmospheric humidity, and the equipment you use. The faster you dry the seeds, the less likely they will be to succumb to pathogens. The lower the humidity of the air in which the seeds are placed, the faster the seeds will dry. Seeds dry quickly at first, then more slowly as their moisture content nears that of the air around them.

PACKAGING SEEDS FOR DRYING

Small coin envelopes work very well for holding seeds while they dry, better than standard paper letter envelopes, which often have small holes at their corners through which tiny seeds can spill out. Another option is to wrap small seeds tightly in pieces of paper towel and secure the seams with tape. Whether you use envelopes or paper towels, package only a few seeds together. If you have a lot of seeds, you'll have to use a lot of envelopes. Be sure you carefully label each envelope or paper towel with the date and the crop or species name (and the variety name if there is one). Note that although plastic bags and glass jars work well for storing dried seeds, they are not good container choices for seeds during the drying process.

DEHUMIDIFIER DRYING

If the air in your home is damp, as is often the case in coastal states and in the South, set up your seeds in a small room with a dehumidifier. You can use an inexpensive hygrometer and a thermometer to estimate how dry seeds are by tracking the relative humidity and maximum and minimum temperatures in the drying room. Measure the relative humidity and temperature daily.

Leave the seeds in the drying area for at least a week to be certain their moisture has come into equilibrium with that in the air. Ideally, follow this rule of thumb: The sum of the relative humidity and the storage temperature in degrees Fahrenheit must not exceed 100, as long as the temperature is less than 50°F (10 °C). For example, if the relative humidity in your drying room is 40 per cent and the temperature is 40°F (4 °C), then the sum is 80 (40 + 40), which is less than 100, so your seeds should dry just fine.

Most orthodox seeds will dry to an ideal 6 to 8 per cent moisture if they're held at 38 to 40°F (3.5 to 4.5 °C) and 30 to 35 per cent relative humidity. Seeds of recalcitrant species like those of the oaks will dry to their appropriate moisture content if held at 38 to 40°F (3.5 to 4 °C) and about 40 per cent relative humidity.

As a homeowner, you don't have the wherewithal to purchase expensive moisture meters to be absolutely sure of having perfectly dried seeds, but these schemes will get you into the ballpark. The graph on page 52 illustrates the relationship between the relative humidity of the drying room and the approximate moisture content of the seeds.

You can easily see that keeping the relative humidity below about 40 per cent will reduce the seed moisture to below about 8 per cent, although more exact percentages depend upon species and seed lot. The table below gives some specific examples of the expected moisture content of vegetable-crop seeds when dried at 40°F (4.5 °C) at 45 per cent relative humidity.

USE OF A DEHUMIDIFIER DRYER

1. Use a dehumidifier dryer to dry a small volume of air in a drying room or other limited space. The humidity of this room will depend on its size, the ambient humidity of the area and the efficiency of the dehumidifier.
2. Place the seeds which have been packaged in labelled porous containers in the drying area.
3. Do not stack the bags too closely and use open racks in a drying room or cabinet with a fan to allow air to circulate.
4. Measure the relative humidity and maximum and minimum temperatures of the room daily.
5. Leave the seeds in the drying area until the moisture content is predicted to be in the range required for storage.

Equipment

- Dehumidifier dryer or cabinet
- Envelopes
- Porous boxes or containers
- Maximum minimum thermometer
- Aspirated hygrometer
- Racks
- Coarse balance
- Spatula/spoons

USE OF SILICA GEL

1. Use deep blue silica gel in an enclosed space, such as a desiccator, can or glass jar with an air-tight seal.
2. Place either the silica gel or the seeds in a porous bag.
3. Use a weight of silica gel equal to the weight of seeds for rapid drying.
4. Place the required weight of silica gel in a desiccator, jar or can with the correct weight of seeds. When using cans or jars make sure that the silica gel is in close proximity to, but not touching, the seeds.
5. Place the containers in a room held at approximately 15 °C whilst the seeds are drying.
6. Change the silica gel daily or when the colour changes from deep blue to pale blue or pink.
7. Heat the pale blue or pink silica gel in an oven above 100 °C until it turns deep blue again, when it is ready for re-use. Store in an air-tight container.
8. Leave the seeds with fresh changes of silica gel in the container until the moisture content is predicted to be in the range required for storage.

Notes and Examples

Desiccators are useful but expensive. Cans and jars with air-tight seals are cheaper and adequate and also allow separation of seeds of different accessions. Silica gel is cheap and easy to use. It can be made into permanent packages of fixed weight in fine cloth bags. These can be heated in the oven without damage and are quicker and easier to handle than loose silica gel.

The silica gel is used to reduce the relative humidity of the atmosphere surrounding the seeds. The seeds then dry by equilibration with their surroundings.

Equipment

- Silica gel
- Desiccators, air-tight cans or bottles
- Forced draught oven from 50 - 200 °C
- Coarse balance
- Spatula/spoons

ACCRATELY DETERMINE THE MOISTURE CONTENT

1. When you consider, from experience, that the seeds are dry enough, remove a sub-sample from each accession and carry out a moisture content determination as explained in Section III.
2. When the moisture content is between 3-7 per cent for most species, go to packaging (Section VI). Low moisture contents are detrimental

to the viability of the seeds of some crops. These seeds should be treated with care and not dried to low moisture levels, *e.g.* soyabean should not be dried below 8 per cent moisture content.

3. If the moisture content is not low enough, continue to dry for a further period as described above. When the moisture content is predicted to be in the correct range for seed storage, carry out an experimental determination as described in Section III.

Notes and Examples

At this point it is necessary to do an experimental determination of moisture content to ensure that predictions of storage life of this accession are as accurate as possible. Make sure that seeds which have been dried are kept in the drying area or moisture-proof containers whilst tests are carried out to prevent absorption of moisture from the surroundings.

Equipment

- Grinder
- Heat resistant dishes with covers
- Analytical balance
- Forced draught oven
- Desiccator
- Silica gel

Tongs and oven cloth

OVEN DRYING

If you live in a fairly humid area and have no dehumidifier, you can try drying seeds in your oven instead. This is easy, works well, and is safer for the seeds than is drying them in direct sunlight. Note that this method is primarily for larger seeds such as those of squash and beans. Tiny seeds such as those of cabbage and carrot seeds should dry sufficiently without heating in an oven.

To prepare seeds for oven drying, do not put the seeds into envelopes or folded paper towels. Instead, spread the seeds in a thin layer on a cookie sheet and place the cookie sheet in the oven for 24 hours at 100°F (38 °C). Stir the seeds once or twice during this time to be sure all sides are exposed to drying air, and that should do it. Once the seeds are dry, you can package them in containers for storage. Some pinecones may require a higher temperature, 130°F (55 °C) or more; we note exceptions like this in the plant entries in part 2.

USING SILICA GEL

Following a dry-air treatment with a silica-gel treatment provides extra insurance that your seeds will be dried well enough. This is advisable if you live in the mid-Atlantic or southern states, or in other areas with very high relative humidity, or in general if you want to be absolutely sure your seeds are ready for storage. Silica gel is widely available at photosupply stores or by mail order from some garden seed companies and is very easy to use. The best bead size for drying seeds is $^{1}/_{16}$•- to $^{1}/_{2}$-inch diameter.

To set up seeds to dry with silica gel, first measure out an amount of the deep blue silica gel equal to the weight of the seeds. Place either the silica gel or the seeds in a porous bag, then place both in an enclosed space such as a glass jar with an airtight seal. Make sure the silica gel is not touching the seeds. Place the containers in a room held at 59°F (15 °C) for drying. You'll need to replace the used silica gel with fresh, either daily or when the colour turns from deep blue to pale blue or pink (which indicates that the gel has absorbed its fill of moisture).

Allow the seeds to sit, changing the silica gel as needed, for a few weeks to be sure they have dried properly. Larger seed lots and bigger seeds require longer drying times. Some folks use powdered milk as an alternative drying agent to silica gel, but it is perhaps one-tenth as effective. Use it as you would the gel. Powdered milk can't be redried very well, so dispose of it.

REUSING SILICA GEL

You can reuse silica gel over and over again as long as you dry it out between each use. To do so, heat used silica gel in a warm oven to drive off the moisture. Set the oven at a temperature above 212°F (100 °C) but below 275°F (135 °C). Place the gel in a thick-walled Pyrex dish in a layer no more than an inch deep. Stir the gel a few times during the drying process; 1 quart (1.9 pounds) of gel should take about 2 hours to dry. You'll know it's dry when it has turned deep blue again. Alternatively, heat the gel in a microwave oven at a medium or medium-high setting for 3 to 5 minutes. If it is still not dry at the end of the cycle, stir it and heat it again for the same period. It may take about 10 minutes to dry a pound. Store the dried gel in an airtight container for future use.

STAY OUT OF THE SUN

Avoid placing seeds in direct sunlight while they are drying. The strong ultraviolet light may damage seed embryos, affecting long-term seed viability.

Should you dry the seeds that you've collected from your garden? In most cases, yes. If you plan to store the seeds for a while before planting, begin drying them as soon as you can after you clean them. However, if you intend to plant the seeds soon after cleaning, take care not to dry them, because this may induce dormancy that does not exist in some fresh seeds.

SEED PROCESSING EQUIPMENT

There are three elements, air, temperature and humidity, that if engineered correctly will result in precision seed drying. Air, its quality, temperature, and movement, have an enormous impact on seed conditioning and productivity. We have discovered that many seed drying problems are caused by an unbalanced drying system. In other words, for every cubic foot of air exhausted an equal amount must be replaced.

The simple solution is the exchange of makeup air with fresh air. Most importantly, the introduction of make-up air eliminates air contamination, exchanges stale air or humidity, eliminates un-proportional air, and maintains proper air stratification for proper seed drying conditions.

SEED DRYING FEATURES

- STS chamber dryers are designed and engineered for precision of temperature, humidity and air flow for drying all varieties of seed.
- Seed drying must be as versatile as possible. Yet, the products' temperature, humidity and air flow have to be accurate. If they are not, under drying or over drying can occur. The STS dryers are designed to provide this accuracy.
- The STS dryer cabinet/chamber models trap air and humidi1y with precise calibrated temperature and discharge them at a given rate based on seed density and seed moisture content.
- The operator has the control to increase or decrease the amount of air velocity in the front portion of the dryer plenum. This control allows the operator to increase or discharge excess amounts of air as needed with regards to drying load capacity.
- Each drying box or chamber has its own damper adjustment. Either to balance air velocity or close off a section.
- Each drying box or chamber section can be calibrated with air velocity different from the other damper section or chambers at anytime.
- The STS electric and gas model heating units are both versatile in air flow to provide a more detailed velocity requirement. The Gas Model, however, uses speed control motors and bleeder dampers to adjust air speed and pressure.

SEED DRYER HEATING COMPONENTS

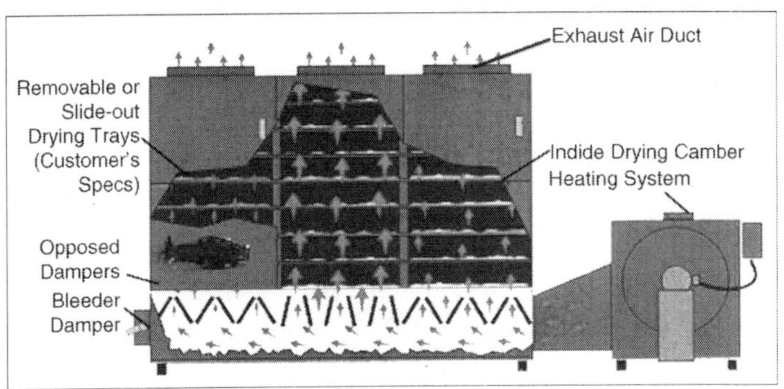

Modulating Controller - Gas Models

STS Dryer systems use a modulating controller to keep the temperature band as tight (narrow) as possible to eliminate heat swing.

The modulating controller senses the temperature of the discharged air and proportions the gas input to load demand. When the temperature is below the controller setting, the burner will modulate to high fire. As the air temperature approaches the controller setting the modulating gas valve will close in small increments until the control point temperature is maintained. As the discharge air temperature falls below the control point setting, a reverse cycle takes place opening the modulating gas valve trying to maintain at least a 3-50F heat swing. Again, as tight of a heat band as possible is maintained.

Electric Models

All electric model heating units are versatile to heat settings. Our electric air flow is calibrated by a variable speed controller Also available is a motor switch control for two speed setting. The heat is controlled by staggering heat elements to keep a consistency of calibrated heat throughout the drying regime. This is to provide a gentle cycle at all times. The dryer can keep temperature consistent to 1 degree heat swing with mild (tight band) heat swing cycling.

PROVEN SEED DRYING

Through our experience we have found that air flow in relation to moisture content in the seed has a great advantage if dried gently and evenly The air releasing excess moisture from the seed at a slow discharge in a stratified chamber will keep the tissue of the seed plump, and won't allow inconsistency throughout the seed bed. This will allow the seed to reach the desired moisture content; but Will provide plump, vigorous and long term viable seed. This is also an important drying process to have dried plump seed to get a precision upgrade when separating seed for its final stages.

8

Seed-Borne Disease

INTRODUCTION

Recent increases in the production and sale of organic seed has heightened the scrutiny of organic seed quality and in particular brought attention to concerns of seed-borne disease contamination. Seed-borne diseases are pathogens such as bacteria, fungus, or viruses, that live on the surface or interior of seed and have the potential to spread disease to the subsequent crop.

Conventional seeds are often treated or coated with chemical fungicides to kill pathogens, a practice not allowed in organic production. Organic seed companies must instead practice careful monitoring and management of seed crops to prevent disease and utilize organically approved treatments to clean infected seed. It is always prudent to seek high-quality seed and work with a reputable company. However, seed-borne diseases are not ubiquitous in all crops, and high concern should be placed primarily on specific diseases known to be seed-borne that pose a risk in your growing region. The use of organic seed does not pose any higher risk than conventional seed if high-quality seed is used.

The risk of seed-borne disease infection varies widely by crop, disease, and location. Many diseases will only become a problem if grown in a region or environment conducive to the disease. Commonly diseases present on seed may also be soil-borne or air-borne and the ultimate fate of the crop may be as dependent on the variety resistance and crop management practices as on the presence of seed-borne innoculum.

There are additionally many microorganisms present on seeds that have no known negative effects and some feel may hold potential positive effects, although there is no current research documentation. It is still important to start with high-quality, clean seed. Seeds of Change tests for a host of common seed-borne diseases on brassicas, tomatoes, and peppers and is working with a Ph.D. plant pathologist in our seed production and quality control programme. In select instances the spread of specific pathogens from seed may introduce the disease to the system with devastating effects. Such is the case of

Watermelon Fruit Blotch, a bacteria that can be seed- or soil-borne and difficult to manage once in the system; particularly in the warm, humid southern region of the United States. The disease is very difficult and expensive to test for and for this reason most companies, including Seeds of Change, require a signed waiver with the purchase of watermelon seed even though it is not a problem in most growing regions. Seeds of Change only grows watermelon seed in regions with no known Watermelon Fruit Blotch pressure.

Disease pressure is often regionally based and in some instances seed use is regulated by government authorities. In these regions intensified agricultural production and high regional economic dependence on a specific crop justifies this regional management approach. Such is the case with bean seed planted in Idaho. All seed must pass stringent disease testing before planting. Most of Seeds of Change's bean seed, including all varieties offered in the Bulk Seed Catalog, are grown in Idaho and produced from certified, disease-tested seed. Lettuce seed in the Salinas Valley of California is also monitored and must pass stringent tests for Lettuce Mosaic Virus prior to planting as this virus is quickly spread by leaf hoppers and the intensive lettuce production in the region poses a high economic risk. This disease is not usually a problem outside of this region. Potatoes in certain states are also certified by government agencies as meeting disease standards and clean-handling practices.

DISEASE MANAGEMENT IN SEED CROPS

In the Seeds of Change Quality Control Programme we work with our seed growers to avoid disease in seed production, we carefully inspect and clean our seed at our own certified organic seed cleaning facility, and we test many of our seeds for specific diseases of concern. One of the most important factors in avoiding disease in seed crops is producing in regions with minimal disease pressure. Many of the Seeds of Change growers are located in western regions where dry summers aid in minimizing disease pressure and allow harvesting seed under dry, late-summer/fall conditions. Additionally we work with our growers and professional breeders to select several crops for disease resistance.

Careful field management in seed production can significantly mitigate seed-borne diseases in seed crops. Many of the same practices that prevent disease in field crops are used to prevent disease in seed crops. Disease promoting conditions are avoided by not using overhead irrigation under moist conditions, using drip irrigation, and cutting water to allow seed crops to dry before harvesting. Spacing plants adequately and orienting rows with the prevailing winds aids in increasing air flow—an important factor in minimizing disease. Rotating crops is important to avoid disease build-up in the soil. Of course, starting with clean, disease-free seed for planting stock is also crucial. Additionally, seed growers must use care in harvesting and cleaning seed to avoid post-harvest infection.

Seeds of Change works with our seed growers to select varieties for disease resistance. In this programme disease is intentionally allowed to build up in an isolated field for breeding and selection purposes. Seed crops are selected under the disease conditions for resistance to the disease. In this case care is taken to ensure the stock seed is not harboring seed-borne diseases when used as planting stock for seed production.

Many of Seeds of Change beet varieties have been selected in this manner for resistance to Rhizoctonia, a bacteria that causes scarring and lesions on beet roots. Many of the lettuce varieties offered have also been selected for resistance to Downy Mildew and Sclerotinia. Seeds of Change growers regularly practice rouging (selection) in seed production fields by removing diseased plants from the field.

Over 1,500 microorganisms have been shown to be related to seeds of all types. However only a small percentage of these organisms pose a threat of disease to the following crop. The majority of common diseases are not seed-borne, but initiate from the local environment. As most diseases are regionally based it is valuable to check with local extension services and growers to identify which seed-borne diseases pose a threat in your ecosystem and then place extra care in sourcing varieties of these crops.

If there are crops of particular concern, ask your seed company for recommendations and information about their disease management programme. And always use good organic disease management practices in the field.

CAUSE

Black rot is caused by a bacteria, *Xanthomonas campestris* pv. *campestris*, that can infect most crucifer crops at any growth stage. This disease is difficult for growers to manage and is considered the most serious disease of crucifer crops worldwide.

The disease can cause significant yield losses when warm, humid conditions follow periods of rainy weather during early crop development. Late infections can provide a wound for other rot organisms to enter and cause significant damage during storage.

Fig. Black rot Infected Cabbage.

SYMPTOMS

Symptoms of black rot vary considerably depending on the host, cultivar, plant age and environmental conditions. The bacteria can enter plants through natural openings and wounds caused by mechanical injury on roots and leaves. Seedborne bacteria infect the emerging seedlings through pores on the margin of the cotyledons and then spread systemically through the seedling.

Infected seedlings grown in the greenhouse under cool conditions (below 15–18 °C) frequently do not show any symptoms of the disease. When infected seedlings are transplanted to the field and temperatures rise to 25–35 °C during periods of high relative humidity (80–100 per cent), they become stunted with dead spots on the cotyledons and will eventually wilt, and die. In regions with temperate climates (where temperatures remain cool), disease symptoms on infected seedlings may not always be obvious or appear severe. Infected seedlings grown under cool conditions may ooze bacteria from pores and lesions, which then serve as a source of the pathogen for neighbouring plants.

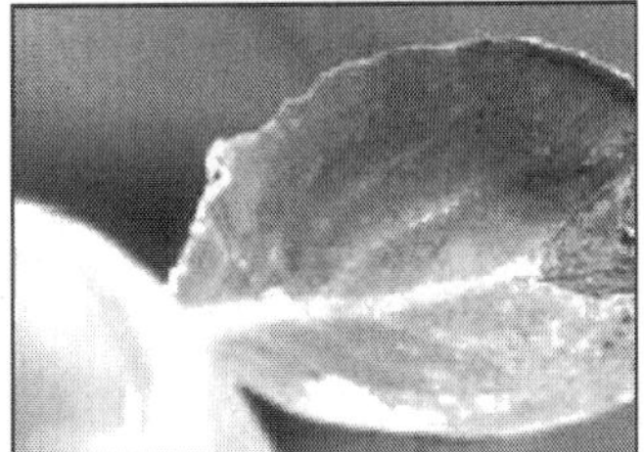

Fig. Young Cabbage leaf with V-shaped Lesion Characteristic of black rot Symptoms.

On older plants, the disease symptoms often appear as yellow or dead tissue at the edges of leaves, similar to tip burn, except the lesion frequently progress into a V-shape with the base of the V usually directed along a vein. Close inspection of infected leaves and stems may reveal black veins running through the infected tissue from which the disease gets its name. Lesions on leaves can expand down towards the base of the leaf causing the leaf to wilt and die.

The bacteria produce a sticky polysaccharide called xanthan that eventually plugs the vascular tissue inside the veins causing them to collapse and turn black. The tissue above the plugged, collapsed xylem eventually turns yellow, wilts and dies. During hot humid environmental conditions, the bacteria can move from the leaf into the stem through the xylem.

Once inside the stem, the bacteria can move up or down to other parts of the plant including the roots. Systemically infected plants may produce chlorotic areas anywhere on the leaf. Severely infected leafy cole crops such as kale and cauliflower tend to shed their leaves from the bottom up leaving only a tuft of distorted leaves separated from the root system by a scarred barren stem. Symptoms on cauliflower often appear as black flecks or scorched leaf margins. The curds of infected cauliflower heads often become blackened.

Fig. Black rot Symptoms Appear as dead Tissue at the tips of (a) Kale, (b) Cauliflower, and (c) Cabbage Leaves. Note the V-Shaped Lesion Progressing from the tip along the vein of the black rot Infected Cabbage leaf.

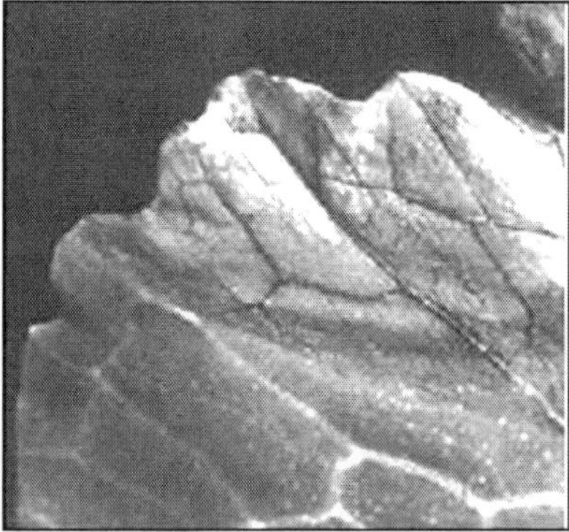

Fig. Black Rotting Veins Running through Black Rot Lesion on tip of Cauliflower Leaf.

Foliar symptoms may not be visible on infected root crops such as rutabaga and radish but blackened vascular tissue can appear inside the edible root tissue rendering the plants unmarketable. Although some infected plants may appear healthy, cutting across infected stems will reveal characteristic blackened vascular tissue. This is a simple method of determining the presence of the disease.

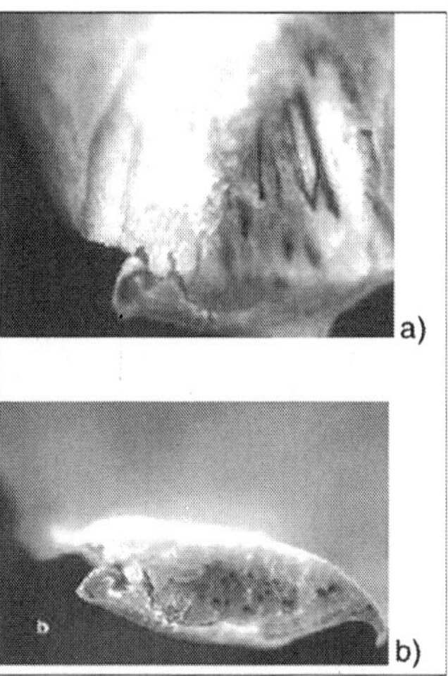

Fig. Cross Section of the base of a Black rot Infected Cabbage (a) Stem and (b) Leaf Revealing the Black, Collapsed Xylem.

Some symptoms of black rot closely resemble those caused by *Fusarium* yellows, which causes the vascular tissue to turn brown. Most commercial crucifer cultivars are resistant to *Fusarium*.

Fig. Leaf Symptoms of Fusarium Yellows Sometimes Appear Similar to black rot Except the Vascular Tissue turns brown Instead of black.

DISEASE SPREAD

Seed contaminated with black rot bacteria is considered the most important source of the pathogen and significantly contributes to the spread of this disease worldwide. As few as 3 infected seeds per 10,000 (0.03 per cent infected seeds) can result in a black rot epidemic. Seed should be tested and certified to be disease free with less than 1 in 30,000 infected seed.

The organism survives in infected crop tissue left on the soil until the crop tissue rots. However, the bacteria do not survive very long in soil as unprotected free living organisms. The black rot bacteria can also infect and survive on many crucifer weeds. This also contributes to the persistence and spread of the disease. It can grow and multiply on host tissue without infecting or causing disease. Rain splashed bacteria from contaminated plant residue left on the soil or from neighbouring diseased plants is the primary method of disease spread throughout a field. The bacteria enter and exit through water-secreting glands called hydathodes located at the edges and tips of leaves. Hydathodes often produce a drop of water during periods of high humidity early in the morning. The pathogen spreads very quickly when rain droplets contaminated with bacteria splash onto healthy leaves and enter the hydathodes. The bacteria move into the leaf veins through hydathodes and begin to multiply, rot and plug the veins. Contaminated water droplets that exude out of hydathodes of infected leaves can then be rain- splashed to other plants. Black rot is more severe and widespread in fields that receive frequent early morning rains, particularly in May and June. Equipment, people, animals and overhead irrigation can further spread the disease. Insects can also spread the bacteria; however, their contribution to the spread of black rot is limited.

Fig. Hydathodes are Special Glands or Pores at the end of Vascular Tissue on leaves through which Water Exudes and are a Natural Opening for Black Rot Bacteria to Infect.

DISEASE MANAGEMENT

Black rot management begins with the identification of potential disease sources and utilising an Integrated Pest Management (IPM) strategy including host resistance, planting disease free seed, avoiding spreading the disease and proper sanitation.

Sanitation is the main method that reduces, excludes or eliminates the initial sources of disease. General sanitation practices include crop rotation, disinfecting seed, rouging diseased plants, elimination of refuse piles and eradication of alternative hosts.

SEED TREATMENT

Seedborne inoculum significantly contributes to the spread of black rot bacteria. Growers should only plant tested certified seed < 1 infected seed in 30,000 or 0.003 per cent contamination. When the infection level of seed is not known or disease-free seed is not available, seed should be treated to eliminate the bacteria. Growers who purchase transplants should request proof the seedlings were grown from disease-free or treated seed. During transplanting, diseased seedlings should not be planted in the field.

Seed treatments do not always eliminate 100 per cent of the bacteria on or in the seed, and may adversely affect seed germination and vigour. Soaking seeds in hot water at 50 °C for 25–30 min. is the most effective treatment for seedborne blackrot control. Weak seed, seed stored for several years and seed of certain crucifer crops; such as, cauliflower, kohlrabi, kale, rutabaga and summer turnip, may be damaged by hot water treatment; soak for 15 min. at 50 °C only.

The effect of the hot water seed treatments on every variety of each individual crucifer crop has not been investigated. Growers are encouraged to treat a small portion of seed and plant in pots to determine the effect of the seed treatment on germination and vigour, prior to treating the entire seed lot.

AVOID DISEASE SPREAD

Use new seed trays each year to avoid contaminating this year's crop with residual black rot bacteria from the previous year. If purchasing new trays each year is not economically feasible, used trays can be sterilised with steam, boiling water or chemical disinfectants to eliminate potential contamination. Destroy infected seed trays immediately to prevent disease spread to other seedling trays.

Avoid soaking crates or bundles of transplant seedlings in tubs of water before transplanting. The black rot bacteria can spread from diseased to healthy seedlings by infecting leaf scars and wounds on roots when soaked in water.

Black rot bacteria can contaminate the surface of clothing, equipment, tools and water sources. Reducing seeding rates and densities to promote good air circulation, facilitating the quick drying of plants, timing irrigation when plants will dry quickly and restricting field activities until later in the day when fields are dry will help reduce disease spread. Working in diseased fields last will also avoid disease spread from infected to non-infected fields. Wash and disinfect equipment before moving from one field to another.

FIELD SELECTION

Field selection is very important due to the distance the pathogen can spread. Whenever possible, select fields as far away from fields grown to crucifer crops the previous year. Select fields that are well drained and will not receive run-off water from areas or fields where crucifers have been grown previously. Well drained, light soils are best for crucifer production because they can be worked early in the season and facilitate earlier planting of transplants. Planting early can help avoid disease because environmental conditions are usually not conducive for the development and spread of black rot bacteria.

CROP ROTATION

Planting disease-free, treated seed or seedling transplants does not necessarily ensure a disease free crop in the field. Crop rotation is also an important management tool. Black rot bacteria can survive in infected crop tissue in soil until the crop tissue breaks down and rots. The time required for crucifer crop debris to rot varies between regions depending on the temperature, amount of soil moisture and soil type. For example, in the states of Georgia and Washington, which experience long, warm summers, it has been estimated that free-living bacteria can survive in infested soil for about 60 days, and up to 615 days in infested host debris. The bacteria can survive longer in soil during cool, wet seasons than during hot, dry seasons. In Ontario, a 3-year rotation is recommended.

WEED CONTROL

Black rot bacteria can infect and survive on many crucifer weeds including bird rape (*Brassica campestris*), Indian mustard (*B. juncea*), black mustard (*B. nigra*), shepherd's purse (*Capsella bursa-pastoris*), globe-podded hoary cress (*Cardaria pubescens*), pepper grass (*Lepidium densiflore*) and wild radish (*Raphanus raphanistrum*). Disease symptoms on weeds vary from small yellow V-shaped lesions on leaf margins to no visible symptoms. The pathogen can spread up to 30 m from infected plants (including weed hosts) to healthy plants. The pathogen not only infects and spreads from weeds to cruciferous crops, it can also survive on weed seeds and can grow and multiply on weed leaves without infecting or causing disease. Good weed control within fields will aid disease management; however, careful attention to weed control in ditches and along fencerows is also important.

INSECT CONTROL

The crucifer flea beetle (*Phyllotreta cruciferae*) can transmit black rot bacteria from infected plants to healthy ones; however, their importance in the spread of the disease is limited. Wounds caused by insects provide an entry point for the disease to infect plants during heavy dews or periods of rain. Insect control will help reduce the spread and severity of disease.

CULL PILE MANAGEMENT

Infected refuse or cull piles left in the field, provides an excellent source of the black rot bacteria. Fresh cull piles left near fields can result in severe disease epidemics during the growing season. Prepare cole crops for market away from fields, and immediately chop and bury the diseased tissue cut from plants.

RESISTANT VARIETIES

The development of crop varieties with disease resistance or tolerance to black rot has been the focus of many cole crop breeding programmes worldwide. Resistance to black rot was first identified in the Japanese cabbage cultivar,

Early Fuji. Today, many crucifer hybrids with black rot tolerance are available for both fresh and processing commercial production.

CHEMICAL CONTROL

Soil fumigation can significantly reduce black rot bacteria. Soil fumigation is expensive and alternative methods for managing plant pathogenic bacteria are needed. For more information on chemical control options refer to OMAFRA Publication 363, *Vegetable Production Recommendations*.

CROP NUTRITION

The effect of plant nutrient management on the susceptibility of host crops to black rot infection is not fully understood. A balanced nutrient programme may reduce the susceptibility of plants to disease infection.

Excess nitrogen promotes lush vegetative growth and may increase plant susceptibility. Micronutrients may also be involved with the disease defence mechanisms of crucifer crops.

DAYLILY RUST

INTRODUCTION

Daylilies (*Hemerocallis* sp.) are one of the top-selling herbaceous perennials in North America. As members of the Liliaceae family, daylilies produce showy lily-like flowers that bloom in clusters, for one day before senescing. Daylilies are native to the Old World from central Europe to China and Japan.

Fig. Flower of 'Pardon Me'.

Most selections available to North American gardeners are cultivated hybrids of original species. Daylilies are most often propagated by crown divisions in spring or fall. These divisions are transplanted into the field or containers for further growth until sale. Daylily cultivation and breeding has become exceptionally popular, with the formation of strong national and provincial/state chapters across North America. Originating in Asia, daylily rust (*Puccinia hemerocallidis*) was first discovered within North America in the southern United States during 2000.

The disease quickly spread and was found in various locations throughout the United States and in Canada during 2001. Daylily rust threatens the production and cultivation of daylilies in North America. Proper identification and understanding the biology of the disease is critical for successful disease management.

SYMPTOMS

Daylily rust is caused by a fungus, *Puccinia hemerocallidis*. This fungus will only infect and colonize green, live host tissue (Figure). It can infect leaves and scapes (leafless, flower stems) of daylily plants but not roots or crowns.

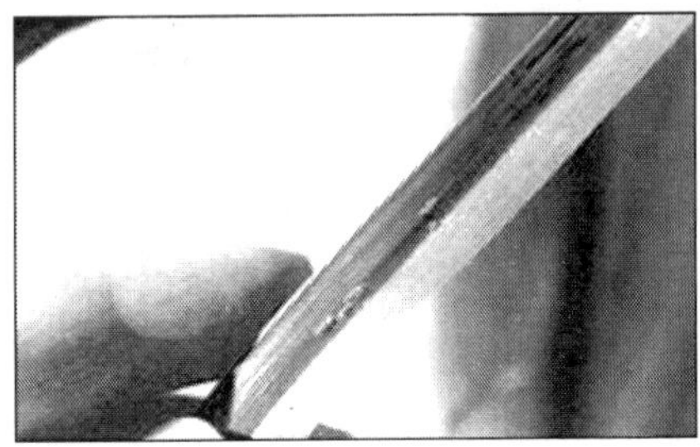

Fig. The Spores of Daylily rust can be easily rubbed off onto Fingers.

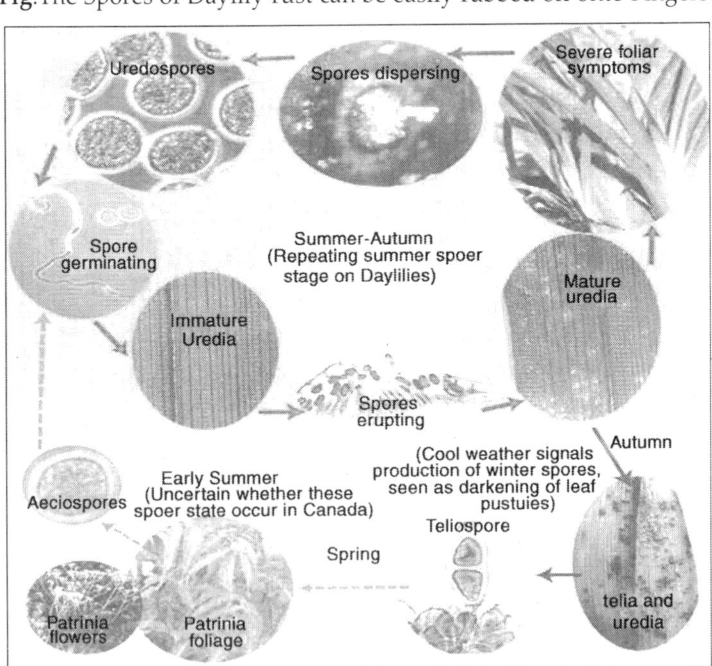

Fig. Lifecycle of Daylily Rust *(Puccinia Hemerocallidis)*. The Alternate hose, *Patrinia,* is shown in the Lower Left. The Dotted Green Arrows Indicate the Uncertain role of *Patrinia* in the daylily rust life Cycle in Ontario

Under conditions favourable for disease development, symptoms appear 3-7 days after infection. Symptoms on very susceptible varieties appear as small, yellow-orange, oval-shaped pustules. The chlorotic areas between the pustules often coalesce and become quite pronounced on more susceptible varieties severe foliar symptoms). On less susceptible varieties, pustules are not as numerous and are surrounded by tan-brown, dead leaf tissue. The pustules contain hundreds of rusty coloured summer spores (uredospores). These spores can easily be transported by wind or rubbed off onto clothing, boots or tools.

It is these summer spores that cause repeat infections on neighbouring daylily leaves and scapes throughout late summer and into autumn.

The fungus requires living green tissue to continue to grow and produce the rust-coloured summer spores. In the autumn, before leaves begin to senesce or die naturally, new infections will produce dark brown to black pustules that contain the resting or winter spores.

BIOLOGY AND LIFE CYCLE

The biology of the fungus *P. hemerocallidis* is complex and requires 2 different plant host species and 5 different rust spore types to complete a life cycle. In the spring, the dark winter spores (teliospores) germinate and produce another set of spores, which can only infect the alternate host,*Patrinia* spp. On *Patrinia*, 2 more spore stages are found with their own distinct symptoms (Bergeron, 2004).

In summer, spores produced from *Patrinia* can infect daylilies. These infections result in yellow spots called uredia which produce the repeating summer spores (uredospores), that re-infect daylilies.

Several cycles of re-infection can occur weekly if conditions for rust infection and development are favourable. With cooler temperatures and leaf senescence, the uredia stop production of uredospores and begin to produce the dark, winter spores called teliospores. During this transition, both teliospores and uredospores can be found in the same pustules, which begin to darken as more teliospores are produced.

The masses of teliospores (called telia) overwinter, and germinate in the spring to continue the life cycle. Again, the telial stage of this fungus is likely not required for the disease to continue on into the next growing season. The fungus may be able to survive on varieties that maintain green leaf tissue during the winter, on plants brought indoors, or on plants overwintering in a minimum-heated greenhouse.

ALTERNATE HOST

The alternate host for daylily rust is the herbaceous perennial *Patrinia* spp. *Patrinia* is found in the Valerianaceae family and is sometimes referred to as Elvis-eyes or Golden Valerian.*Patrinia* spp. have small yellow flowering clusters that grow from a clump of palmately-lobed leaves, and are often used in lightly or partly shaded gardens.

The plants are native to Asia, but several species including *P. gibbosa*, *P. triloba*, *P. villosa* and *P. rupstris* are sold in the United States as well as in warmer regions of Canada. *Patrinia* is not a common perennial in North American gardens and likely poses little threat to the spread of this disease in the landscape. However, there is still some uncertainty about the role of *Patrinia* in the disease cycle of daylily rust in North America. There is very little information regarding the survival of daylily rust in northern temperate regions such as Canada. However, the disease may not require the alternate host, *Patrinia* spp., since the fungus may be able to survive on varieties which maintain green leaf tissue during the winter, on plants brought indoors, or on plants overwintering in a minimum-heated greenhouse.

Environmental Conditions for Disease Development

In Ontario, daylily rust is not usually observed until the latter part of summer and early autumn. Studies have shown that the optimal temperatures for summer spore germination can occur at 22-24°C under high humidity (although germination can occur from 7-34°C). Summer spores do not germinate in cold (<4°C) or extremely warm temperatures (>36°C) and this disease is not severe during hot, dry or cold conditions. A minimum of 5-6 hours of continuous leaf wetness is required for spore germination and leaf infection. In addition, summer spore germination decreases with high light intensity. Under conditions favourable for disease, hundreds to thousands of summer spores from each infected leaf can be produced quickly and spread rapidly, making this disease a serious threat to daylilies. This phase of the disease cycle can repeat itself many times during periods of warm weather with rain or dew periods, resulting in disease epidemics.

VARIETY SUSCEPTIBILITY

Daylily rust does not necessarily kill plants but can affect plant vigour, susceptibility to other pests and marketability. Differences in cultivar susceptibility to daylily rust have been observed in experiments conducted at the University of Georgia and the University of Guelph. However, a limited number of cultivars have been evaluated and more research into cultivar susceptibility and resistance is required. Varieties known to be very susceptible (many pustules containing numerous summer spores) include: 'Buttercup', 'Catherine Woodbury', 'Cherry Cheeks', 'Colonel Scarborough', 'Couble', 'Imperial Guard', 'Irish Ice', 'Ming Toy', 'Pardon Me', 'Karie Ann', 'Lemon Yellow', 'Little Gypsy Vagabond', 'Pandora's Box', 'Quannah' and 'Russian Rhapsody'. Moderately susceptible (fewer pustules frequently surrounded by dead brown-tan coloured leaf tissue and containing fewer summer spores) varieties include: 'Butterflake', 'Condon', 'Crystal Tide', 'Gerturde', 'Happy Returns', 'Prelude to Love', 'Joan Senior', 'Pandora's Box', 'Rosy Returns', 'Star

Struck', 'Stella D'Oro', 'Summer Wine', 'Wilson's Yellow' (and*Hemerocallis fulva*). Low susceptibility (little to no infection) varieties include: 'Butterscotch Ruffles', 'Holy Spirit', 'Mac the Knife' and 'Yangtze'.

DIAGNOSING DAYLILY RUST

If you suspect your plants have been infected with daylily rust, reduce the chances of spreading this disease by placing a large, clear plastic bag over the symptomatic foliage and obtain leaf samples from within the confines of the bag. Close the bag tightly around the base of the plant and remove as much infected foliage as possible inside the bag. Take the samples indoors where foliage can be examined more closely.

Raised, orange pustules can be seen with the naked eye on both the upper and lower leaf surfaces but may be more prevalent on the lower leaf surfaces of some varieties. Pustules are more easily viewed with a 10-20x magnification hand lens. Immature pustules are elliptical, raised bumps with a waxy sheen and a yellow-orange colour.

Mature pustules have a protruding mass of orange, powdery spores. The spores can be easily rubbed off onto fingers. Submit a sample to the local Pest Diagnostic Clinic for conclusive identification.

DISEASE MANAGEMENT

- Avoid growing *Patrinia* spp. in nurseries or gardens where daylilies are grown. Spores produced on *Patrinia* spp. are thought to be infective to daylilies. If the different hosts are in close proximity, then the chances of infection are greater.
- Scout daylily plants frequently during the later weeks of summer, particularly after rain, when conditions are favourable for dew formation or during a prolonged period of cloudy days with day time temperatures around 22-24 °C.
- Select and grow the least susceptible varieties in regions where daily rust has been found.
- Avoid overhead irrigation. To minimize leaf wetness periods, direct irrigation to the soil surface and not the leaf canopy. Divide daylilies every 3-5 years and keep plants well spaced to facilitate leaf drying after rain or irrigation. If possible, irrigate plants in the morning to allow for quick drying of foliage during the day, rather than watering in the evening.
- If daylily rust is confirmed, cover infected plants with a large, clear plastic bag and remove all leaves and scapes from the infected and surrounding plants as close to the ground as possible. Composting of diseased tissue is not recommended at this time since it is not known how long the summer spores will survive under composting conditions

or whether they may blow out of the compost pile and infect healthy plants nearby. Diseased tissue should be placed in a plastic bag and promptly removed from the area. Use disposable rubber gloves and wash clothing after working with or around diseased plants.

- Use registered fungicides to help protect healthy plants from rust infection. Fungicide applications should begin in early summer and may terminate when daytime temperatures do not exceed 7 °C. If plants become infected but are not yet producing powdery spore masses, systemic fungicides may be able to cure the infections and prevent pustule development.
- In the autumn, cut plants back to remove all green foliage. Do not mulch as mulching infected plants may protect rust pustules that have gone unnoticed and possibly allow them to survive the winter.

DISEASES OF ASPARAGUS

INTRODUCTION

Three diseases caused by fungi affect asparagus in Ontario. Fusarium root and crown rot is caused by a soil-borne fungus, while asparagus rust and Botrytis blight are caused by fungi whose spores are primarily carried about by wind. The life cycles and characteristics of these fungi will be discussed in this Factsheet to develop an understanding of how they may be recognized and controlled.

FUSARIUM ROOT AND CROWN ROT

Description

The major disease problem of asparagus is caused by two species of fungi called *Fusarium*. *Fusariummoniliforme* causes decay of storage roots, stems and crowns. This fungal species is present in all agricultural soils and infects corn, grasses, and other monocotyledonous plants as well as asparagus. *Fusarium oxysporum* f. sp. *asparagi* causes root rot and seedling blight, and may also plug the water conducting vessels, causing wilting of spears and fern.

These species of soil-dwelling fungi are very prolific, long-lived, and capable of growing as saprophytes, feeding on decaying asparagus residues and soil organic matter. They colonize old roots and crowns, invading directly through root tips or through wounds from implements, cutting tools, or insect feeding. Asparagus plants which are under stress are more susceptible to infection than those which are growing vigorously.

Affected spears may shrivel and rot in spring before or after emergence. Infected crowns have hollow, rotted feeder and storage roots. When crown and stem tissue is sliced open, a reddish-brown discoloration is visible. Symptoms on fern includes stunting, yellow to brown discoloration of one or more stalks per crown, and fewer stalks per crown. Affected crowns decline in vigour and die. Such damage is scattered throughout the field and increases until the stand is too sparse to harvest economically.

Fig. Yellow Discoloration of Asparagus Fern Infected by *Fusarium*.

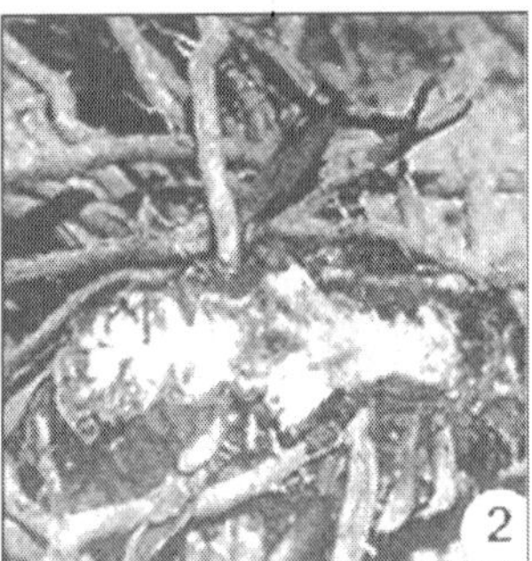

Fig. Cross-section of Asparagus Crown Infected by *Fusarium*.

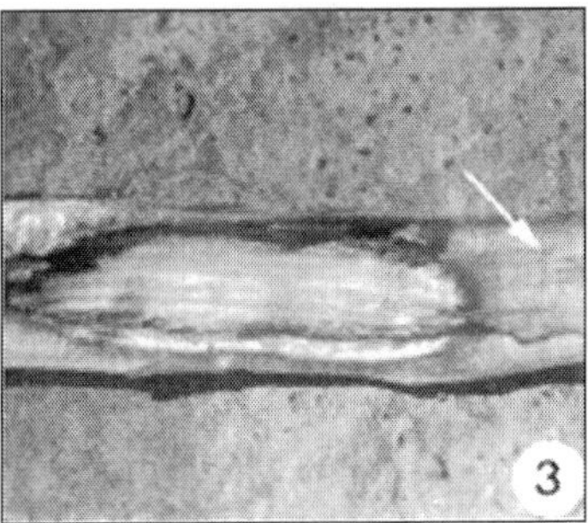

Fig. Cross-section of Asparagus Spear Infected by *Fusarium*.

CONTROL

Because *Fusarium* species are present in most agricultural soils, the diseases are almost impossible to avoid. Control is therefore directed at minimizing infection early in the life of crowns, and at maintaining a vigorous, long-lived asparagus stand by careful management.

Suggested management practices include:

1. Avoid replanting in land which previously grew asparagus. *Fusarium* builds up to extremely high levels during the long period of asparagus culture in the field, and survives many years in that soil even after the crop is removed.
2. Soil fumigation with a suitable material will reduce infection in crown nurseries or direct-seeded asparagus.
3. Use treated seed. Untreated seed should be disinfested with 1 part sodium hypochlorite bleach in 5 parts water for two hours. Rinse seed in fresh water and dry before planting.
4. Use only vigorous, one year old crowns and encourage proper handling procedures for transplanting. Weak crowns are highly susceptible to infection.
5. Minimize stress in young and established plantings. Stresses which weaken asparagus crowns and thus promote *Fusarium* diseases are:
   - Overextending the harvest season, or picking before crowns are well established;
   - Insect and disease damage;
   - Weed competition;
   - Mechanical wounding from tillage equipment, cutting knives, etc.;
   - Poor soil drainage;
   - Acidic soil pH;
   - Drought;
   - Injury from misapplication of pesticides, fertilizers, and soil ammendments;
   - Low fertility; and
   - Soil compaction.
6. Use resistant varieties. At present, no varieties recommended in Ontario are resistant. However, lines and varieties being developed in the University of Guelph breeding programme are being screened for *Fusarium* resistance.

ASPARAGUS RUST

Description and Life Cycle

Puccinia asparagi, which causes asparagus rust, has a complicated life cycle consisting of several stages, all of which occur on asparagus. Some members of the onion family, such as cooking onions and chives, are also susceptible. There are no alternate hosts such as is common with other *Puccinia* rusts on wheat and oats. The fungus overwinters as teliospores on asparagus debris. Teliospores germinate in spring, producing small, basidiospores which are blown onto emerging spears and cause infection. Later, from April until July, small upraised, light-green, oval lesions (patches) called aecia occur on the lower portion of the infected fern stalks.

As the aecial lesions turn creamy orange in colour, aeciospores are released and re-infect asparagus fern during several hours of continuous leaf wetness. Twelve to fourteen days after re-infection, upraised tan blisters called uredia appear on asparagus stalks and foliage. Uredia break open to expose masses of rusty-coloured spores called uredia spores, after which the disease is named. Urediospores repeatedly re-infect asparagus from June until September. Warm weather with heavy dew, fog, or light rainfall enhances rust development. Late in summer, telia develop, producing black teliospores, completing the yearly life cycle.

DAMAGE

The fungus develops in the tissue of asparagus fern and drains the plant of vital nutrients. The foliage then dries out and falls prematurely, further reducing the production and storage of food for the following crop. Successive years of infection by rust will weaken asparagus crowns.

CONTROL

1. Cultural control involves breaking the continuity of the rust life cycle. Inoculum from overwintered teliospores in fern residue may be spread from wild or neighbouring asparagus, uncut spears, young or seedling plantations, or volunteer seedlings in the asparagus crop. It is important to:
   - Destroy wild and volunteer asparagus;
   - Clean cut all fields during harvest; and
   - Locate nurseries and young plantings remotely from established fields if possible.
2. Use resistant asparagus varieties. Viking varieties have some degree of tolerance to rust, but will become infected under conditions of favourable environment and abundant inoculum. Breeding and variety trials at the University of Guelph are including rust resistance as an important criterion in the development of new varieties.
3. Control with fungicides is presently limited to the chemical zineb. Spraying should begin when the aecial or initial uredial stages are evident on the fern - in May on young plantings, in late June in harvested plantings. Because zineb is a protectant fungicide, adequate coverage*before* rust appears and throughout the season is essential to achieve control.

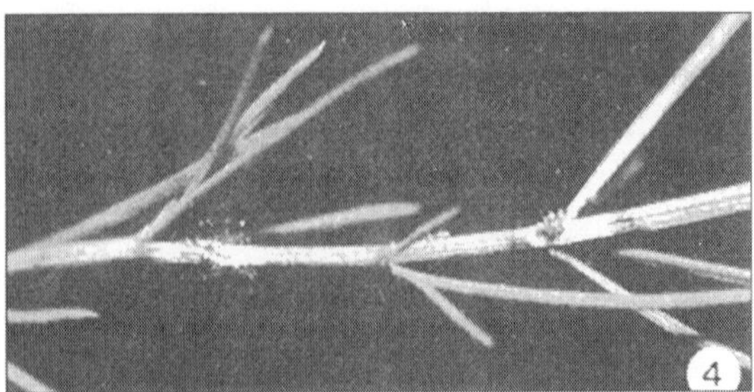

Fig. *Botrytis* Lesions on Asparagus Fern Branches. Note the Tufts of Fruiting Bodies Produced by the Fungus in these Lesions.

Fig. Extensive Blighting of the Lower fern Canopy Caused by *Botrytis*.

Fig. Aecial Lesion of Asparagus Rust. Note Oval to Elliptical Shaped, Upraised Lesion.

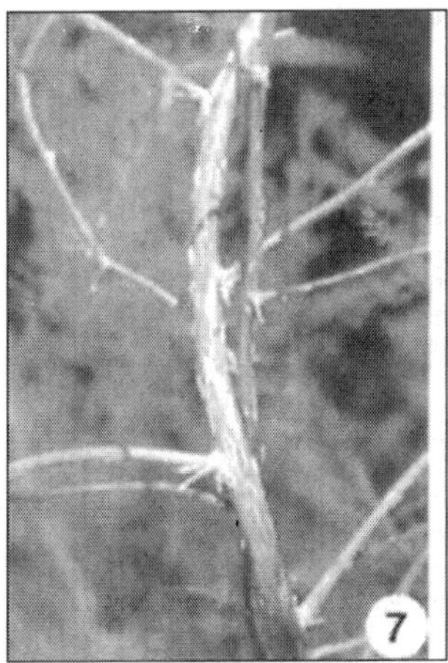

Fig. Uredial Lesions of Asparagus Rust. Note Tan Blisters which have Opened, Exposing the Rusty Coloured Spore Mass.

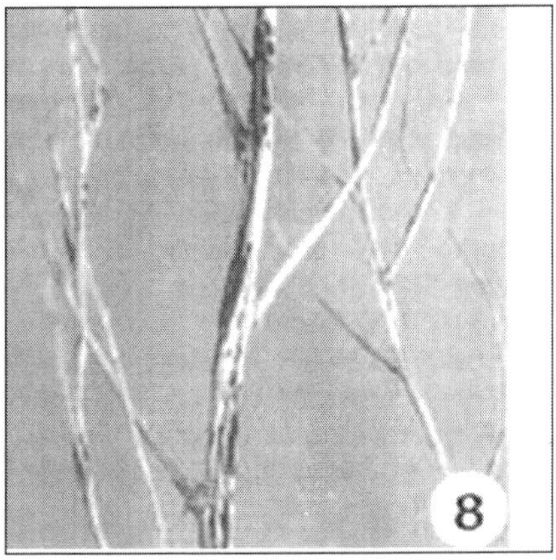

Fig. Telial Lesions of Asparagus Rust. Note Black Teliospores on Overwintering Asparagus Fern.

- Botrytis Blight
- Description and Life Cycle

This disease, caused by the fungus *Botrytis cinerea*, occurs during summer, causing browning of the lower fern canopy (Figure). *Botrytis* progresses most rapidly during hot, moist weather when the fern does not dry adequately. This fungus attacks many other crops, such as strawberries and other small fruit, potatoes, beans, and other vegetables, and many ornamentals.

The disease begins on senescing (dying) flowers or injured fern. Botrytis spores are spread by wind and rain within the dense fern canopy. Individual lesions are tan with dark brown borders, often surrounded by a fellow halo (Figure). When wet weather persists, newly-emerged spears may be completely blighted, turning brown or black, and often covered with grey, fuzzy, spore-bearing fungal growth.

Control

Zineb will also give some control of *Botrytis* when used regularly for rust control.

Viruses

Though several viruses have been identified in asparagus in Europe and Western USA, they have not yet been reported in Ontario.

Pesticide Application

As fern growth becomes dense, more water and higher rates of pesticides must be used to obtain adequate spray coverage. Good coverage is essential, especially for protective fungicides like zineb. Drop nozzles are recommended. Up-to-date pesticide recommendations are available for commercial growers in OMAFRA Publication 363, *Vegetable Production Recommendations*. Homeowners are referred to OMAFRA Publication 64, *Insect and Disease Control in the Home Garden.*

In conclusion, as a perennial crop, asparagus is a long-term investment. It will pay to monitor pests regularly throughout the harvest and growing seasons. Only then can problems be detected and controlled before they cause long-term economic damage.

9

Seed Testing Management

INTRODUCTION

Tests of the quality and other characteristics of seed need to be made at several stages during its progress from the parent tree to the seed bed. Tests for maturity and soundness of seed before and during collection in the forest and are usually associated with surveys of the abundance of the seed crop; they are intended to ensure that both the quantity and the quality of seed justify the effort and cost of collection. Several tests may be required at the seed processing depot, after extraction and cleaning and before the seeds are despatched to the nurseries or placed in storage.

Testing of moisture content is essential for many species before they are stored for a long time, but is usually unnecessary if the seed is to be sown immediately in the nursery; on the other hand, partial drying may be needed before despatch of seed to a distant forest station, to reduce the risk of deterioration in transit, and any deliberate drying operation must be accompanied by the testing of moisture content. Testing of germination or viability should be repeated at the end of the storage period if this is more than a few months and, in the case of long-term storage for conservation of genetic resources, testing should be done at intervals throughout the storage period (Ellis *et al.* 1980).

Efficiency and success in raising plants in the nursery and in their subsequent establishment in forest plantations depend to a great extent on the quality of the seeds used. It follows that foresters, nurserymen, seed dealers and others need accurate estimates of the quality of the seeds in which they deal or which form the basis of their afforestation projects. This is particularly important when seeds are bought and sold, or move between countries in international trade.

The example of estimating seed demand which is shown in Table uses estimates of the number of germinated seeds per kg of Pinus kesiya and Tectona grandis which are averages for the species. In practice, the number varies considerably from year to year and from seed lot to seed lot, so that it is necessary for the forester to have an accurate estimate of germination for every individual seed lot which he receives, if he is to fulfil planting targets while at the same time avoiding the waste of valuable seed caused by sowing too much. Similarly the nurseryman must have a good estimate of germination if he is to adjust sowing rates to produce the optimum spacing of seedlings in the seedbed.

Justice (1972) has defined the following objectives for developing rules for seed testing:

a. To provide methods by which the quality of seed samples can be determined accurately;
b. To prescribe methods by which seed analysts working in different laboratories in different countries throughout the world can obtain uniform results;
c. To relate the laboratory results, insofar as possible, to planting value,
d. To complete the tests within the shortest period of time possible, commensurate with the above-mentioned objectives;
e. To perform the tests in the most economical manner.

For the practising forester, the most important object of seed testing is the provision of an accurate estimate of the capacity of a given seedlot to produce healthy, vigorous plants suitable for field planting. In the present context, seed "quality" refers to the physiological vigour of the seed rather than its genetic quality.

The essence of good seed testing is the application of reliable standard methods of examination to ensure that uniform and reproducible results are obtained. Standardization has been greatly facilitated through the adoption by a number of countries of the International Rules for Seed Testing formulated by the International Seed Testing Association (ISTA). ISTA was founded in 1921, drew up its first set of Rules in 1931 and made major revisions of the Rules in 1953, 1966 and 1976. In the early stages ISTA's primary concern was for agricultural seeds, but trees and shrubs have gradually assumed greater importance and the 1976 Rules (ISTA 1976) give (firm) "prescriptions" or (more tentative) "suggestions" as to testing methods appropriate for 61 different genera of these, compared with 26 in the 1953 Rules. The Rules are published in English, French and German.

Although tree species and genera from the tropics and southern hemisphere (notably eucalypts) have begun to figure in the ISTA lists, the balance is still overwhelmingly weighted towards north temperate species. Thus important species such as Tectona grandis, Pinus patula, P. oocarpa and P. kesiya are listed, but only in the list of "suggested" (non- prescriptive) seed testing methods, while a recent survey (ISTA 1981 a) has shown that 20 other important tropical species are not included at all; they include Cupressus lusitanica, Gmelina arborea, Cordia alliodoraand the genera Albizzia, Araucaria, Casuarina, Swietenia, Terminalia and Triplochiton. Their omission simply reflects the lack of reliable information from research on the best methods of testing these species and genera.

ISTA testing methods involve the maintenance of controlled laboratory conditions and the use of some items of equipment which are relatively expensive. They are therefore well suited to large, well-equipped seed laboratories but are impracticable for small forest stations which have no laboratory and must carry out germination tests in the nursery or the office. Lack of equipment is no reason to omit seed testing altogether. For example simple germination tests carried out in the nursery can produce results which are perfectly satisfactory for local use. But precise details of method should always accompany results whenever this is not standardized, so that the seed user can interpret the results in relation to his own nursery conditions.

An up-to-date guide to the range of equipment available for seed testing is contained in the "Survey of Equipment and Supplies" published by ISTA (ISTA 1982). It lists equipment by type, use, model specification and supplier. The earlier directory, "Equipment and supplies for collecting, processing, storing and testing forest tree seed" includes seed testing equipment and gives the addresses of user laboratories as well as suppliers.

Van der Burg *et al.* (1983) give a description of how a seed testing station in tropical or subtropical areas could be established. Two alternatives are described: Seedlab 2000, that can test about 2000 samples per year and Seedlab 5000 that can test at least 5000 samples per year. Directives and general considerations concerning the staffing, the organization of the work, the lay-out of the building, and the equipment needed are given. Forty-six figures and two tables give an impression of the equipment and administrative forms used. Equipment which in the experience of the authors has been found suitable for the work is recommended and detailed descriptions and company addresses are mentioned. Some equipment that is not commercially available is described and plans for construction are included. A list of books and journals for a basic seed testing library is appended.

The tests which may be required are purity, authenticity, seed weight, germination, indirect testing of viability, moisture content, and seed health and damage. A prerequisite for all testing is good sampling, which is described in the following section.

SAMPLING

When seed is tested, the sample on which the test is made has to be representative of the whole. No matter how accurate the technical work in the test, the results can only show the quality of the sample submitted for analysis (Aldhous 1972). Therefore, every effort must be made to ensure that the submitted sample accurately reflects the composition of the whole seed lot. This is equally applicable, whether a whole series of tests is to be made in a laboratory, or a simple nursery test carried out to determine number of germinated seeds per kg of uncleaned seed. Time spent testing carelessly drawn samples may be time wasted.

Completely homogeneous seed lots would be easy to sample, but they do not exist. Common sense measures can be taken to reduce heterogeneity as much as possible, for example seed lots of the same species should not be mixed if they come from different origins (provenances), differ greatly in age or, in the case of the same introduced provenances, if they come from plantations growing on very different site types in the introducing country. But a seed lot may be heterogeneous even if collected from a homogeneous stand. Paul (1972) has given the simple example of pine seed transported over bad roads in the back of a landrover.

The bumps and vibrations to which it is subjected will cause a settling out of different types of material. All the small and empty seeds and other light rubbish will eventually reach the surface layers and a sample taken from the top alone will give a completely false impression of the potential performance of the seeds as a whole.

MIXING

If the seed lot is a small one consisting of only a few kg of seed, it is possible to improve its homogeneity by thorough mixing of the whole lot before taking a sample.

In the case of very large seed lots transported and stored in many different containers, mixing of the whole seed lot is impracticable. Instead, a number of different samples are taken, as described on pp. 194–195, and it is these samples which are thoroughly mixed together to form a homogeneous "composite" sample for testing. The methods of mixing which are quoted below from Paul (1972), are equally applicable to small entire seed lots or to composite samples.

"*Methods of mixing. There are two simple ways in which seed may be thoroughly mixed*:

i. Mixing with the aid of a mechanical divider. Mechanical dividers are used to reduce the size of seed lots or samples by repetitive halving. Their operation is described below. The divider has the added advantage that it can also be used to mix seed in the following manner:
    1. Pass entire seed lot through divider.
    2. Take the two equal portions and feed them simultaneously into the divider.
    3. Repeat operation No. 2.
    4. Repeat operation No. 2 once more.
    5. Take the two equal portions and pour them simultaneously into the storage container.

ii. *Handmixing*: It is surprising how difficult it can be to achieve a homogeneous seed lot using this method. The entire seed lot must be spread out on a sheet of paper or some other suitable smooth surface and mixed by scooping the seed from side to side and from top to bottom. After thoroughly mixing, spread the seed out evenly and divide the lot into four equal parts. Place these separately in four containers and with the aid of an assistant pour them simultaneously into the storage container. Repeat the spreading, quartering and pouring process another two times."

USE OF SEED TRIERS

For a large seed lot contained in a number of different containers, seed triers are used to take small sub samples from different parts of the lot. All of these "primary" samples are mixed together to form a "composite" or bulk sample, which is later reduced in size by successive dividings until it is small enough to form the "working" sample on which the various tests are carried out.

As described by Turnbull (1975 d), a seed trier is a probe, long enough to reach all areas of the seed bag and designed to remove an equal volume of seed from each area through which it travels. Most commonly used is a sleeve-type trier which consists of a hollow brass tube inside a closely fitting outer sleeve. The tube and sleeve have open slots in their walls so that, when the tube is turned or slid, the slots in the tube and the sleeve coincide and seeds can flow into the tube; when the tube is given a half turn or slid in the opposite direction, the openings are closed.

Seed triers should have a series of separate compartments and not just a single long tube with several openings. Tubes are available in varying lengths and diameters to cater for different sizes of container and kinds of seeds. Triers known as "thief" triers, with a single open slot, should be avoided, as they can damage the seeds.

Ideally, the primary samples should be distributed proportionally to the volumes of the different parts of the seed lot. For example, if a lot is contained in ten 10-kg containers and ten 20-kg containers, the 20-kg containers should contribute two thirds of the composite sample and the 10-kg containers only one third. For seed lots in containers of uniform size, ISTA has detailed prescriptions as to the number of primary samples appropriate for varying numbers of containers, *e.g.,* for 6 – 30 containers at least one in three containers is to be sampled and never less than five.

Large seeds and others that are not free-flowing are not suitable for sampling by seed trier. They should be sampled by thrusting the hand into the seeds and removing small portions. The hand should be inserted flat with the fingers extended together. The fingers should be kept together as the hand is closed and withdrawn. It is difficult to sample with this method deeper than about 40 cm and containers may have to be partially emptied to facilitate sampling.

REDUCING THE SIZE OF COMPOSITE SAMPLES

Usually the composite or bulk sample for testing needs to be reduced to a working sample of standard weight. Whenever possible the sample should be subdivided by a mechanical divider to reduce any operator bias in the dividing. The non-mechanical halving method is described here for use if a mechanical divider is not available, but it is the least desirable method for most types of seed. Manual methods must, however, be used for seeds which are not free-flowing.

NON-MECHANICAL METHODS OF DIVIDING

Halving Method

The sample is placed on a clean surface and thoroughly mixed by hand; it is then divided into quarters with a sharp-edged spatula and opposite quarters discarded. The process is repeated until a final sample of the approximate weight required is obtained.

Random Cup Method

A series of small cups or thimbles are arranged on a tray in a definite pattern and the sample is poured systematically over this area. The working sample is obtained from randomly selected containers (Thomson and Doyle, 1955). A modification of this method uses a tray divided into an equal number of square compartments, every alternate one of which has no bottom.

MECHANICAL METHODS OF DIVIDING

Most mechanical dividers are designed to split the sample into two approximately equal parts. The working sample is obtained by repeatedly dividing the bulk sample until it reaches the required weight. ISTA recommends three types of dividers as being suitable equipment. They are the conical divider (Boerner type), the soil divider, and the centrifugal divider. They are briefly described as follows:

Boerner Sampler. This divider is made in different sizes. The essential parts consist of a hopper, inverted cone, and a series of baffles directing the seeds into two spouts. The baffles form alternate channels and spaces of equal width. They are arranged in a circle at their summit and are directed inward and downward, the channels leading to one spout and the spaces to an opposite spout. A valve or gate at the base of the hopper controls the seed flow. When the valve is opened the seeds fall by gravity over the inverted cone where they are evenly distributed to the channels and spaces; then, they pass through the spouts into the seed pans below. In Zimbabwe a similar arrangement on a larger scale is used for dividing and mixing large seed lots. From the inverted cone there are 8 outlets which pass the seed into buckets, and a battery of 8 buckets is available for each outlet. The buckets hold 6 kg of pine seed each, so that a seed lot of up to 384 kg can be handled in any one dividing and remixing operation.

Fig. A seed trier.

Fig. Random Cup Divider.

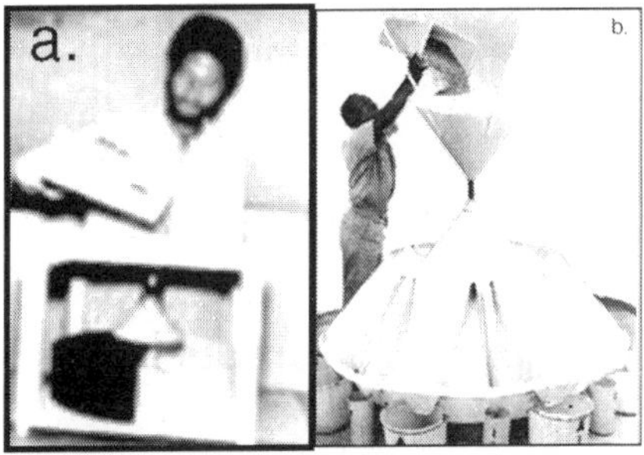

Fig. Inverted Cone Dividers used (A) for Small (B) for Large Quantities of Seed.

Fig Seed Dividers (A) Boerner (B) Gamet.

Fig. An Opaque Glass Screen, Illuminated from below, used for Purity Tests and Determinations of the Number of Seeds per kg.

Fig. Two Types of Weighing Scales used in Seed Laboratories.

Soil Divider

A simple divider, built on the same principle as the conical divider, is the so-called soil divider or riffle divider. The channels are here arranged in a straight row instead of a circle as in the conical divider. The soil divider consists of a hopper with attached channels or ducts, a frame to hold the hopper, two receiving pans and a pouring pan. In using the divider the seed is scattered fairly evenly in a pouring pan and poured in at approximately equal rates along the entire length of the hopper. The divider is suitable for large-seeded and chaffy species, but suitable types for small-seeded species can also be made.

Gamet Divider

The Gamet divider makes use of centrifugal force to mix and scatter the seeds over the dividing surface. The seeds flow downward from a hopper onto a shallow rubber cup below. Upon rotation of the rubber cup by an electric motor the seeds are thrown out by centrifugal force and fall downward. The circle or area where the seeds fall is equally divided into two parts by a sharp-edged, stationary baffle so that one-half of the seeds fall in one spout and the other half in the other spout. In using this divider care must be exercised in dividing very small samples as it is possible that a majority of seeds may be thrown out in one spout. Hardin *et al.* (1965) compared the effectiveness of the Boerner and Gamet dividers and concluded that if used correctly, the Gamet divider yields a slightly more accurate subsample.

WEIGHT OF SAMPLE

The weight of the working sample depends on the seed size of the species in question. The ISTA rules are intended to provide a minimum of 2500 seeds for all except the very large-seeded species, for which a minimum of 500 seeds is prescribed. These quantities are considered sufficient for most of the usual tests (purity, authenticity, seed weight, germination or viability), but an additional 10 g of most species is needed if moisture content is to be determined. Among tropical species listed by ISTA the range of prescribed minimum size of working sample is from 2 g for Eucalyptus deglupta, which averages 4,000 viable seeds per gm of "seed plus chaff" to 1 kg forTectona grandis which averages 2,000 fruits per kg. When seed is despatched to an independent seed laboratory for testing, ISTA recommends that the weight of the submitted sample be double that of the working sample.

PURITY ANALYSIS

Tree seed samples can contain impurities such as weed seeds, seeds of other tree species, detached seed structures, leaf particles and other material. The object of purity analysis is to determine the composition by weight of the sample being tested. To do this the sample is separated into component parts. When purity analysis is done, it is the first test to be carried out because subsequent tests are made only on the pure seed component.

Pure seed refers to the species under consideration and in addition to mature, undamaged seed includes: undersized shrivelled, immature and germinated seeds, provided they can be definitely identified as the species under consideration; and pieces resulting from breakage that are more than one-half their original size. Seeds of Leguminosae and Coniferae with the seed coats entirely removed are regarded as inert matter. Other components of the sample may include Other seeds of all species except that under test and Inert matter. Inert matter consists of pieces of broken or damaged seeds less than half the original size, wings of coniferous species, leguminous and coniferous seeds with the seedcoat entirely removed and other matter such as fragments of leaves, twigs, stones, soil.

In Coniferae (except Chamaecyparis,Cupressus and Thuja) any remaining seed wings not already removed in cleaning should be detached and classed as inert matter.

The working sample containing all the impurities is weighed and then the pure seed is removed and weighed separately. ISTA rules prescribe that weighing be done in gms to the minimum number of decimal places necessary to calculate the percentage of its component parts to one decimal place.

The percentage of pure seed is calculated as follows:

$$\text{Purity}\,\% = \frac{\text{Weight of "pure" seed}}{\text{Total weight of sample}} \times 100$$

When a laboratory is testing a large number of samples for purity, the purity analyst usually examines and separates seed samples on a "workboard" placed on top of a desk or table. The workboard can be adjusted to a desired height, usually 7 – 15 cm above table height. The analyst's equipment should permit him to do his work with a minimum of effort and time, and also with a minimum of eye strain (Justice 1972).

Equipment required for making purity analysis is as follows: forceps and spatula for handling, separating, and pushing seeds over a surface; wide-field hand lens of 5X to 7X magnification; reading magnifiers relatively free of curvature and distortion, wide-field stereoscopic microscope with a range of magnifications from 10X to 75X; scale with capacity up to 1000 g and sensitive to 0.5 g; torsion balance with capacity of 120 g and accurate to 0.01 g; rapid-acting chemical balance accurate to 1 mg; small seed containers to hold separations; and seed blower. An assortment of small sieves can help in eliminating many impurities.

Other methods reported by Turnbull (1975 d) as reducing hand labour are specific gravity separation by differences in density, separations by electrostatic charges, a vibrating table working on the basis of seed weight, surface texture and shape, and an X-ray radiograph technique.

The pure seed component from the purity analysis can be subsampled for the germination test, as well as for the determination of seed weight. Since the germination test is based on pure seed, it can be readily seen that the purity analysis and germination test are complementary to each other. The production potential of a seed lot can be determined only when the purity analysis and the germination tests are considered together.

SEED WEIGHT

Measurement of seed weight is made on the pure seed component separated by the purity test. Weight is normally expressed as the weight of 1000 pure seeds. This figure can be readily converted to number of pure seeds per g or per kg as required.

Weight may be determined simply by counting out 1000 seeds and weighing them, but the use of several smaller samples enables the analyst to estimate variation within the sample. ISTA prescribes 8 replicates of 100 seeds each, from which the standard deviation and coefficient of variation may be calculated, as well as the mean. If the co-efficient of variation is less than 4, the mean is accepted, but, if it is more, 8 additional replicates are prescribed, a new standard deviation for the 16 replicates calculated and any replicate that diverges from the mean by more than twice the standard deviation is discarded before calculating the final mean for the sample.

The counting of seeds for weight may be done by hand or with the assistance of counting boards or vacuum or electronic counter. The following description of the essential features of seed counting devices is taken mainly from Magini.

Counting boards are useful for relatively large seeds of a regular shape, such as Pinus pinea or some leguminous species. The counter consists of two boards, a stationary top board with perforations and a thin solid board below, which serves as a false bottom.

In operating the counter, a quantity of seed is placed at the corner, the board tilted and moved so that one seed goes into each perforation. The board can then be emptied and the known number of seeds from it weighed. Alternatively, if the seeds are being counted in preparation for a germination test, the counter can be placed on the substrate and the bottom is withdrawn permitting the seed to fall through at a uniform spacing. In some types there is a second set of perforations in the bottom board. By slipping the top section about one centimetre, the seed falls through both boards onto the germination bed. In other types, the counting apparatus is perforated by holes too small to allow the seeds to pass through but suitable to allow a flow of air to suck the seeds into position.

In connection with tests on soil trays, a simple counting perforated board can also be employed. The board is first placed over the tray. The seeds are placed in the perforations and the board is lifted leaving the seeds on the tray.

Vacuum counters can be used with seeds of various sizes and shapes. For very small and irregularly shaped seeds some difficulties may occur. They are not suitable for very large seeds, such as many nuts.

A vacuum counting device has three essential parts: a vacuum system including pipes; counting heads; a quick release valve. A small house vacuum cleaner with a home-made head may be used, but special power units are generally preferable (1/4 to 1/2 hp electric motors). Perforated heads, with evenly spaced holes (generally 100) to house the seed, hold the seed by suction. When suction is switched off, the seeds are released for weighing or onto the substrate for germination testing. Counting plates or heads are of a square, rectangular or circular form, according to the shape of the substrate employed. Espacement and diameter of holes should be proportionate to the seed size. Heads can be of plastic, chromium, brass or aluminium. The face plate should be relatively opaque and of a colour which contrasts with the seeds being counted, thus making it easier to check that all holes are properly filled. A good valve is needed to allow a quick release of seed, as well as to avoid sucking up too large a proportion of the lightest seeds. The risk of picking up an excessive proportion of light, empty seeds in the counted sample may be reduced by inverting the head so that the holes are upwards and then pouring a large sample of seeds over it. Operation of vacuum counters with plastic heads may be adversely affected by static electricity.

In the modern types of electronic seed counters a vibrating feed bowl moves seeds past an electric eye in single file. The machine can be programmed to count a certain sample size, then cut itself off (Bonner 1974).

Reservations have been expressed about the accuracy of vacuum and electronic counters on the grounds that they select a biassed sample. Manual counting boards, combined with careful random sampling, are likely to be more accurate, although considerably slower.

The 1000 pure seed weight can be converted to seeds per g or per kg as follows:

$$\text{Number of seeds per g} = \frac{1000}{\text{weight in g of 1000 seeds}}$$

or,

$$\text{Number of seeds per kg} = \frac{1000 \times 1000}{\text{weight in g of 1000 seeds}}$$

If the sample counted is other than 1000 seeds, the appropriate formula is:

$$\text{Number of seeds per g} = \frac{\text{No. of seeds in sample}}{\text{weight of sample in g}}$$

or,

$$\text{Number of seeds per kg} = \frac{\text{No. of seeds in sample} \times 1000}{\text{weight of sample in g}}$$

Within a given species full seeds have a higher specific gravity and a higher germination rate than seeds of the same size but empty or partially filled. Large seeds weigh more per seed than small ones of the same specific gravity and, because they contain larger food reserves, they are likely to germinate better and produce initially more vigorous seedlings. Goor and Barney have reported that large seeds of Eucalyptus citriodora had a higher germination rate than medium-sized seeds which in turn had a higher rate than small seeds. Number of pure seeds per unit weight is therefore not in itself a good guide to potential for plant production and must be complemented by germination tests or indirect tests of viability. The effect of seed size on the growth of germinated seedlings of eucalypts usually persists for 8–14 weeks after sowing, thereafter other factors may become more important.

10

Effect of Preservation Technologies and Microbiological Inactivation

INTRODUCTION

Traditionally, the most popular preservation technologies for the reduction of microbial contamination of food, and pathogens in particular, have been the manipulation of the water activity and/or pH, heat treatments, the addition of chemical preservatives, and the control of storage temperature of foods. Lately, and mainly as a result of consumer demand for "fresher products," other technologies are emerging as alternatives for extension of product shelf life (for better quality products) and reduction of pathogenic organisms (for safer products).

The process by which a product is manufactured is one of the factors to be considered when determining if a food needs temperature control for safety. The efficiency of the process is dependent on a number of parameters unique to each technology that will be described briefly in this chapter. To determine the pathogen reduction needed for the food to be safe at room temperature, other factors need to be considered as well, such as water activity and pH of the food, packaging, processing, formulation, and opportunities for post-process contamination.

Inactivation of microorganisms is influenced by a number of microorganism-related factors that are generally independent of the technology itself. These include the type and form of target microorganism; the genus, species,and strain of microorganisms; growth stage;environmental stress selection mechanisms; and sub-lethal injury. Each factor influences the bacterial resistance independently of the apparent inactivation capacity of that particular process. For pasteurization purposes, one is mostly concerned with the inactivation of vegetative cells of disease-producing microorganisms. However, to have a commercially sterile product, the process must control or inactivate any microbial life capable of germinating and growing in the food under normal storage conditions.

VALIDATION OF PROCESSING PARAMETERS

Establishment of traditional thermal processes for foods has been based on two Main Factors:

1. Knowledge of the thermal inactivation kinetics of the most heat-resistant pathogen of concern for each specific food product; and
2. Determination of the nature of heat transfer properties of the food system.

The validity of a thermal process must be confirmed by an inoculated challenge test conducted on the product under actual plant conditions using surrogate microorganisms as biological indicators to mimic pathogens. Thus, the two factors described above, which are well established for thermal processes, should be used for establishing and validating scheduled new thermal processes based on thermal effect on microorganisms, such as microwave heating.

For other preservation processes not based on heat inactivation, key pathogens of concern and non-pathogenic surrogates need to be identified and their significance evaluated. Surrogate microorganisms should be selected from well-known non-pathogenic populations, should mimic the target pathogenic microorganism in growth habits, should not be susceptible to injury, and should not exhibit non-reversible inhibition (thermal or otherwise). Surrogate microorganisms should be genetically stable and exhibit uniform thermal and growth characteristics from batch to batch over several generations. The durability to food and processing parameters should be similar to that of the target organism. Population of surrogates should be constant and maintain stable thermal and growth characteristics from batch to batch. Enumeration of surrogates should be rapid and should utilize inexpensive detection systems that easily differentiate them from natural flora.

Genetic stability of surrogates is desirable to obtain reproducible results. It also is recommended that surrogates do not establish themselves as "spoilage" organisms on equipment or in the production area. The validation process should be designed so that the surrogate exhibits a predictable time-temperature process character profile that correlates to that of the target pathogen. Introduction of system modifications or variables, leading to inaccurate results should be avoided (for example, thermocouple probes changing heating rates, nutrients added to the product for surrogate growth altering viscosity, and so on).

PROCESSING TECHNOLOGIES

WATER ACTIVITY AND PH

The Manipulation of water activity and/or pH is the less complicated of technologies in terms of equipment, expense, and expert personnel needed.

Although it may not reduce the microbial load *per se*, reducing water activity or pH may retard or impede microbial growth. (For a more extended description on how water activity and pH can be used as preservation technologies and for a list of the optimum range pH and water activity for various pathogens of concern.

When changing these characteristics of foods with the intention of safely storing a food at room temperature, those minimum pH and water activity values should be taken as guidance. At different temperatures and for different foods, these ranges may vary. For example, as the temperature moves away from optimum, a higher minimum pH is generally observed.

TECHNOLOGIES BASED ON THERMAL EFFECTS

In Addition to microbial inactivation by conventional methods of heating, microwave and ohmic and inductive heating are also considered to be heat-based processes that can inactivate microorganisms by thermal effects. Microwave and radio frequency heating refers to the use of electromagnetic waves of certain frequencies to generate heat in a material through two mechanisms-dielectric and ionic.

Ohmic heating is defined as the process of passing electric currents through foods or other materials to heat them. Ohmic heating is distinguished from other electrical heating methods by the presence of electrodes contacting the food, frequency, and waveform. Inductive heating is a process wherein electric currents are induced within the food due to oscillating electromagnetic fields generated by electric coils. No data about microbial death kinetics under inductive heating have been published.

For any of these heat-based processes, the magnitude of time/temperature history and the location of the cold points will determine the effect on microorganisms. The effectiveness of these processes also depend on water activity and pH of the product. Although the shape of the inactivation curves is expected to be similar to those in conventional heating, the intricacies of each of the technologies, however, need special attention if this technology is used for microbial inactivation.

For instance, in microwave heating a number of factors influence the location of the cold points, such as the composition, shape, and size of the food, the microwave frequency, and the applicator design. The location of the coldest-point and time/temperature history can be predicted through simulation softwares, and it is expected that food processors may be able to use them in the future.

For determining the kinetics and efficiency of inactivation of microorganisms for these technologies, surrogate/indicator microorganisms could be selected from those traditionally used in thermal processing studies.

HIGH PRESSURE PROCESSING

High pressure processing (HPP), also described as high hydrostatic pressure (HHP) or ultra high pressure (UHP) processing, subjects liquid and solid foods, with or without packaging, to pressures between 100 and 800 MPa. Process temperature during pressure treatment can be specified from below 32 °F (0°C) to above 212 °F (100°C). Commercial exposure times can range from a millisecond pulse to over 20 min. Chemical and microbiological changes in the food generally will be a function of the process temperature and treatment time. The various effects of high hydrostatic pressure on microorganisms can be grouped into cell-envelope-related effects, pressure-induced cellular changes, biochemical aspects, and effects on genetic mechanisms.

HPP acts instantaneously and uniformly throughout a mass of food independent of size, shape, and food composition. Compression will uniformly increase the temperature of foods approximately 5 °F (3°C) per 100 MPa. Compression of foods may shift the pH of the food as a function of imposed pressure and must be determined for each food treatment process. Water activity and pH are among the critical process factors in the inactivation of microbes by HPP.

An increase in food temperature above room temperature, and to a lesser extent, a decrease below room temperature increases the inactivation rate of microorganisms during HPP treatment. Temperatures in the range of 113–122 °F (45–50°C) appear to increase the rate of inactivation of food pathogens and spoilage microbes. Temperatures ranging from 194–230 °F (90–110°C) in conjunction with pressures of 500-700 MPa have been used to inactivate sporeforming bacteria such as *Clostridium botulinum*. Current pressure processes include batch and semi-continuous systems, but no commercial continuous HPP systems are operating.

The Critical process factors in HPP include pressure, time at pressure, time to achieve treatment pressure, decompression time, treatment temperature (including adiabatic heating), initial product temperature, vessel temperature distribution at pressure, product pH, product composition, product water activity, packaging material integrity, and concurrent processing aids. Interestingly, because HPP acts instantaneously and uniformly through a mass of food, package size, shape, and composition are not factors in process determination.

High hydrostatic pressures can cause undesirable structural changes in structurally fragile foods such as strawberries or lettuce (for example, cell deformation and cell membrane damage). Food products that have been brought to market include raw oysters, fruit jellies and jams, fruit juices, pourable salad dressings, raw squid, rice cakes, foie gras, ham, and guacamole.

A Biphasic pressure inactivation curve is frequently encountered for both vegetative bacteria and endospores indicating the residence of a small pressure-resistant sub-population. Tailing phenomena should be investigated carefully in challenge studies. The use of pathogens rather than surrogates for highly infective pathogens may be advised.

The Elimination of spores from low-acid foods presents food-processing and food-safety challenges to the industry. It is well established that bacterial endospores are the most pressure-resistant life forms known. One of the most heat-resistant pathogens, and one of the most lethal to human beings, is *C. botulinum*, primarily types A, B, E, and F. As such, *C. botulinum* heads the list of most pressure-resistant and dangerous organisms faced by HPP. Spore suspensions of strains 17B and Cap 9B tolerated exposures of 30 min to 827 MPa and 167 °F (75°C) (Larkin and Reddy; personal communication; unreferenced). Because some types of spores of *C. botulinum* are capable of surviving even the most extreme pressures and temperatures of HPP, there is no absolute microbial indicator for sterility by HPP. Among the sporeformers of concern, Bacillus cereus has been the most studied because of its facultative anaerobic nature and very low rate of lethality.

Normally, Gram-positive vegetative bacteria are more resistant to environmental stresses, including pressure, than vegetative cells of gram-negative bacteria. Among the pathogenic non-sporeforming gram-positive bacteria, *Listeria monocytogenes* and *Staphyloccocus aureus* are the two most well-studied regarding the use of HPP processing. *Staphyloccocus aureus* appears to have a high resistance to pressure.

There appears to be a wide range of pressure sensitivity among the pathogenic gram-negative bacteria. Patterson and others (1995) have studied a clinical isolate of *Escherichia Coli* O157: H7 That possesses pressure resistance comparable to spores. Some strains of *Salmonella* spp. have demonstrated relatively high levels of pressure resistances. Given these pressure resistances and their importance in food safety, *E. coli* O157:H7 and *Salmonella* spp. are of key concern in the development of effective HPP food treatments. For vegetative bacteria, non-pathogenic *Listeria* innocua is a useful surrogate for the foodborne pathogen, L. monocytogenes. A non-pathogenic strain of Bacillus may be useful as a surrogate for HPP-resistant *E. coli* O157:H7 isolates

Commercial Implications

Current practical operating pressures for commercial HPP food treatment intensifiers and pressure vessels are in the order of 580 MPa (85,000 psi). If this pressure is specified, then the following process times may be considered as first estimates for initial process planning. It must be understood that actual process parameters must be developed from challenge test packs.

Experience with acid foods suggests that shelf-stable (commercially sterile) products, having a water activity close to 1.0, and pH values less than 4.0, can be preserved using a pressure of 580 MPa and a process hold time of 3 min.

This treatment has been shown to inactivate 106 cfu/g of *E. Coli O157*: H7, *Listeria* spp., *Salmonella* spp., or *Staphylococcus* spp. in salsa and apple juice.

Acid foods with pH values between 4.0 and 4.5 can be made commercially sterile using a pressure of 580 MPa and a hold time of 15 min. Products would have an initial temperature of about 71.6 °F (22°C). Shorter hold times are possible if the product is to be refrigerated. Actual hold-time values must be determined from challenge packs and storage studies perhaps twice the length of the intended shelf life of the product.

Low-acid products can be rendered free of pathogens or pasteurized by HPP; however, satisfactory guidelines for hold times at 580 MPa for low-acid food pasteurization have not emerged. For example, the post-package pasteurization of vacuum-packed cured meat products to eliminate *Listeria* spp. represents a useful application of HPP. Ground beef can be pasteurized by HPP to eliminate *E. Coli O157:* H7, *Listeria* spp., *Salmonella* spp., or *Staphylococcus* spp. Much more work is required to develop a suggested hold time at 580 MPa due to the potential for tailing. Changes in product colour and appearance may limit the usefulness of HPP treatment pressures above 200 to 300 MPa.

PULSED ELECTRIC FIELDS

High intensity pulsed electric field (PEF) processing involves the application of pulses of high voltage (typically 20–80 kV/cm) to foods placed between two electrodes. PEF may be applied in the form of exponentially decaying, square wave, bipolar, or oscillatory pulses at ambient, sub-ambient, or slightly above ambient temperature for less than 1 s. Use of PEF can reduce energy usage compared to thermal processes, as less energy is converted into heat, which also reduces detrimental changes to the sensory and physical properties of the food.

To date, PEF has been applied mainly to extend the shelf life of foods. Application of PEF is restricted to food products that can withstand high electric fields, have low electrical conductivity, and do not contain or form bubbles. The particle size of the food in both static and flow treatment modes is a limitation. Also, due to the variations in PEF systems, a method to accurately measure treatment delivery is still needed.

Factors that affect the microbial inactivation with PEF are process factors (electric field intensity, pulse width, treatment time and temperature, and pulse waveshapes), microbial entity factors (type, concentration, and growth stage of microorganism) and media factors (pH, antimicrobials and ionic compounds, conductivity, and medium ionic strength).

Many researchers have studied the effects of pulsed electric fields in microbial inactivation; however, due to the numerous critical process factors and broad experimental conditions used, definite conclusions about specific pathogen reductions cannot be made. Research that provides conclusive data on the PEF inactivation of pathogens of concern is clearly needed. Castro and others (1993) reported a 5-log reduction in bacteria, yeast, and mold counts suspended in orange juice treated with PEF. Zhang and others (1995) achieved a 9-log reduction in *E. coli* suspended in simulated milk ultrafiltrate treated with PEF by applying a converged electric field strength of 70 KV/cm for a short treatment time of 160 s. This processing condition may be adequate for commercial food pasteurization that requires 6- to 7-log reduction cycles (Zhang and others 1995).

However, numerous critical process factors exist and carefully designed studies need to be performed to better understand how these factors affect populations of pathogens of concern. Currently, there is little information on the use of surrogate microorganisms as indicators of pathogenic bacteria when PEF is used as a processing method. Selection of surrogates will require the prior identification of the microorganism of concern in a specific food and PEF system.

The selection of the appropriate surrogate(s) will depend on the type of food, microflora, and process conditions (that is, electric field intensity, number of pulses, treatment time, pulse wave), and should also follow the general guidelines listed in the validation section.

IRRADIATION

Lrradiation of food refers to the process by which food is exposed to enough radiation energy to cause ionization. Ionization can lead to the death of microorganisms due to genetic damage which prevents cellular replication. For the treatment of foods, FDA has approved the use of gamma rays from decaying isotopes of cobalt-60 or cesium 137, x-rays with a maximum energy of five million electron volts (MeV), and electrons with a maximum energy of 10 MeV. An electron volt is the amount of energy acquired by an electron when accelerated by one volt in a vacuum. X-rays are produced when high energy electrons strike a thin metal film. Lethality of irradiation depends on the target (microorganism), condition of the treated item, and environmental factors. Addition or removal of salt or water, time/temperature of the treatment, or oxygen presence are factors that will influence the antimicrobial effect of irradiation.

Irradiation is considered an additive in the U.S. and as such, it needs to be approved by the FDA office of premarket approval for each new application and labeled. Two terms have been used to define the extent of pathogen reduction with irradiation. Radiation pasteurization refers to the destruction of pathogenic, non-spore-forming foodborne bacteria. In radiation pasteurization, medium dose treatments (1 to 10 kGy) reduce microbial populations, including pathogens in foods. Elimination of pathogens on meat, seafood, and poultry by medium dose irradiation has been studied. Sterilization radiation is used for radiation processes that will render the food commercially sterile or for foods that are both sterile and shelf stable. In this last case, sterilization must ensure the elimination of the most resistant pathogen, endospores of *Clostridium botulinum*. In order to achieve this, higher doses (42–71 kGy depending on the product) than the ones currently permitted for foods (up to 10 kGy, except for spices) are needed. Only frozen meats consumed by NASA astronauts have been permitted by FDA to be sterilized through irradiation.They are, however, in the market in other countries.

Lonizing radiation is used as a means of extending the shelf life of produce (Diehl 1995; Thayer and others 1996). FDA has approved the use of ionizing radiation with a range dose 0.3–1 kGy for growth and maturation inhibition. Not much effort has been applied to the control of foodborne pathogens on fresh foods, mainly because most medium and high level doses are not appropriate for produce since they can cause sensory defects (visual, texture, and flavour) and/or accelerated senescence (Thomas 1986; Barkai-Golan 1992). Ionizing irradiation has recently been used to eliminate E.Coli 0157: H7 from apple Juice, and E. Coli 0157: H7 and Salmonellae from seed and sprouts. Doses in the range of <1 to 3 kGy have been shown to reduce or eliminate populations of foodborne pathogens, postharvest spoilage organisms, and other microorganisms on produce (Moy 1983; Urbain 1986; Farkas 1997). Strawberry shelf life can be extended with treatments in the range of 2 to 3 kGy (Sommer and Maxie 1966; Zegota 1988; Marcotte 1992; Diehl 1995). Research conducted since that time suggests that irradiation can be an important treatment to enhance safety of other types of produce.

FDA and the U.S. Department of Agriculture Food Safety and Inspection Service have also approved irradiation to control foodborne pathogens in raw poultry with a dose range of 1.5–3.0 kGy. Recently, the irradiation of raw refrigerated and frozen meat has been approved with maximum doses of 4.7 and 7 kGy, respectively. Radiation doses of 2.5 kGy in beef will result in 6 log reduction of *Campylobacter*, 5 log reduction of *E.Coli* O157:

H7, 3 Log reduction in Salmonellaspp., and 5 log reduction of *Staphylococcus* cells (CAST 1996). Although the potential for consumer infection by pathogens is decreased greatly and shelf life is extended by radiation pasteurization of meat and poultry, the room temperature storage of raw meat products would be highly discouraged.

Other products, such as shell eggs (up to 3 kGy), have recently been approved for irradiation for safety reasons. Shell eggs can be irradiated with the intention of significantly reducing populations of *Salmonella* spp. Reduction levels depend upon radiation dose, initial level of pathogen contamination, or other treatment-related conditions.

The effect of irradiation on microbial populations suggests that it could also be used to decrease pathogensin a product with the intention of allowing microbiologically safe storage at ambient temperature for a specific time. However, the foods currently allowed to be irradiated are very limited. One could envision that in the future a food such as pumpkin pie could be irradiated to allow for safe ambient temperature storage. As with other technologies, organoleptic changes in the food would need to be considered. More important, the effectiveness of the technology will need to be validated for the specific application.

OTHER TECHNOLOGIES

Some of the technologies present greater limitations or are at a development stage that requires extensive further scientific research before they can be commercially used. For instance, high voltage arc discharge (application of discharge voltages through an electrode gap below an aqueous medium) causes electrolysis and highly reactive chemicals. Although microorganisms are inactivated, improved designs need to be developed before consideration for use in food preservation. Likewise, oscillating magnetic fields have been explored for their potential to inactivate microorganisms; however, the results are inconsistent. Data on inactivation of food microorganisms by ultrasound (energy generated by sound waves of 20,000 or more vibrations per second) are scarce, and limitations include the inclusion of particulates and other interfering substances.

Ultraviolet (UV) light is a promising technique, especially in treating water and fruit juices. A 4-log bacterial reduction was obtained for a variety of microorganisms when 400 J/m^2 was applied. Apple cider inoculated with *E. coli* 0157:H7 treated in that manner achieved a 5-log reduction (Worobo 2000). Critical factors include the transmissivity, the geometric configuration of the reactor, the power, wavelength, and physical arrangement of the UV source, the product flow profile, and radiation path length.

SELECTION OF CHALLENGE ORGANISMS

Some pathogens that may be used in challenge studies for various types of foods (Vestergaard 2001). Knowledge of the food formulation and history of the food (for example, association with known illness outbreaks and/or evidence of potential growth) is essential when selecting the appropriate challenge pathogens. For example, *Clostridium botulinum* would be of concern with certain modified atmosphere packaged (MAP) products, and *Staphylococcus aureus* may be of concern in foods with little competitive microflora and in products with reduced a_w.

The ideal organisms for challenge testing are those that have been previously isolated from similar formulations. Additionally, pathogens from known foodborne outbreaks should be included to ensure the formulation is robust enough to inhibit those organisms as well.

Multiple specific strains of the target pathogens should be included in the challenge study. It is typical to challenge a food formulation with a "cocktail" or mixture of multiple strains in order to account for potential strain variation. It is not unusual to have a cocktail of 5 or more strains of each target pathogen in a challenge study. For example, botulinal challenge studies typically include several strains of proteolytic types A and B as well as representative non-proteolytic strains (if appropriate). A single challenge strain with specific well-defined characteristics may be used to screen products similar in nature to those formulations that have been extensively challenged with multiple strains in the past.

Prior to conducting the challenge study, selected strains should be screened for mutual antagonism. Antagonism between certain strains of *Listeria monocytogenes* has been reported, as well as between certain strains of *C. botulinum* and between bacteriocin-producing lactic acid bacteria. If these are not compatible when used as part of a challenge cocktail, then erroneous results may ensue.

It is also important to incubate and prepare the challenge suspension under standardized conditions and format. Shifts in the incubation temperature used to propagate the challenge organisms and the storage temperature of the product have been shown to change the length of the lag period of the challenge study itself (Curiale 1991). Consideration must also be given to adapting the challenge suspension to the environment of the food formulation prior to inoculation. For example, acid-adaptation of *E. Coli* O157: H7 cells or *Salmonellae* cells prior to inoculation can greatly influence their ability to survive when inoculated into an acidic food.

Finally, the use of genetic characterization tools may greatly aid the determination of which strains (if any) used in the challenge study are the most dominant over the length of the study. Tools such as ribotyping and pulsed-field gel electrophoresis can help distinguish among strains, whether inoculated or naturally occurring (Farber and others 2001).

Challenge organisms may also be genetically modified to carry a marker that aids in distinguishing them from other strains and helps in their detection in the food matrix itself. Obviously, the genetically modified organism needs to have physiological characteristics comparable to the wild-type or parent strain.

For certain applications, surrogate microorganisms may be used in challenge studies in place of specific pathogens. For example, it usually is not possible to introduce pathogens into a processing facility; therefore, it is desirable to use surrogate microorganisms in those cases. An ideal surrogate is a strain of the target pathogen that retains all other characteristics except its virulence. In practice, however, many surrogates are closely related to but not necessarily the same species as the target pathogen.

Traditional examples include the use of *Clostridium* sporogenes as a proxy for *Clostridium botulinum* in inoculated pack studies, *Listeriainnocua* as a surrogate for *L. monocytogenes*, and generic strains of *Escherichia coli* as substitutes for *E.Coli* O157:H7. Caution must be used with the latter, however, since generic strains of *E. Coli* do not have the same level of acid resistance as *E. Coli* O157:H7. Therefore, while it may be appropriate to use generic strains of *E. Coli* in a challenge study to assess the impact of a heat process or preservative system in a high-pH food, it would be inappropriate to use these generic strains for evaluating acidic foods.

Generally, Surrogates are selected from a group of well-characterized organisms and have the following desirable attributes (IFT):

- Non-Pathogenic.
- Inactivation characteristics and kinetics that can be used to predict those of the target pathogen.
- Behaviour similar to the target pathogen when exposed to formulation and/or processing parameters (for example, pH stability, temperature sensitivity, and oxygen tolerance).
- Stable and consistent growth characteristics.
- Easily prepared to yield high-density populations.
- Once prepared, population remains stable until utilized.
- Easily enumerated using rapid, sensitive, and inexpensive detection systems.
- Easily differentiated from background microflora.
- Attachment characteristics that mimic those of the target pathogen.
- Genetically stable so that results can be replicated independently of laboratory or time of experiment.
- Will not establish itself as a "spoilage" organism if used in a production area.
- Susceptibility to injury similar to that of the target pathogen.

If a surrogate strain is to be used in a microbiological challenge study, preliminary work should be done to well characterize the strain before use in the study. Characteristics such as those discussed above should be determined and confirmed through preliminary laboratory work to assure that the surrogate strain is suitable for the intended purpose. The use of surrogates should be limited to only those cases where specific pathogens absolutely cannot be used for product or personnel safety reasons.

INOCULUM LEVEL

The inoculum level used in the microbiological challenge study depends on whether the objective of the study is to determine product stability and shelf life or to validate a step in the process designed to reduce microbial numbers. Typically, an inoculum level of between 10^2–10^3 cells/g of product is used to ascertain the microbiological stability of a formulation. Higher inoculum levels may be appropriate for other products. Depending on the product formulation, some of the inoculum may die off initially before adapting to the environment.

If too low of an inoculum level is used, the incorrect assumption could be made that the product is stable when it is not. Conversely, if the inoculum level is too high for this purpose, the preservation system or hurdles to growth may be overwhelmed by the inappropriate inoculum size, leading to the incorrect conclusion that the formulation is not stable. When validating a process lethality step such as heat processing, high pressure processing, or irradiation, however, it is usually necessary to use a high inoculum level (for example, 10^6–10^7 cells/g of product) to demonstrate the extent of reduction in challenge organisms. For example, in the United States, juice processors are now required to demonstrate a 5 log reduction of relevant hazardous microorganisms in their products (5 D performance standard).

These log-reduction validation protocols usually require the use of plating methods. In order to measure this level of reduction within the statistical limits of the enumeration method, the inoculum level must be at least 10^6 CFU/g.

INOCULUM PREPARATION AND METHOD OF INOCULATION

The Preparation of the inoculum to be used in microbiological challenge testing is an important component of the overall protocol. Typically, for vegetative cells, 18–24 h cultures revived from refrigerated broth cultures or slants or from cultures frozen in glycerol are used. The challenge cultures should be grown in media and under conditions suitable for optimal growth of the specific challenge culture.

In some studies, specific challenge organisms may be adapted to certain conditions. Such adaptation will be tailored to the specific food. For example, *E. Coli* O157:H7 May be acid adapted with the appropriate acidulant prior to use in the challenge studies on acidic products. Bacterial spore suspensions may be stored in water under refrigeration or frozen in glycerol. Spore suspensions should be diluted in sterile water and heat-shocked immediately prior to inoculation. Spores of *C. botulinum* should be washed thoroughly prior to use to ensure that no free botulinal toxin is carried over into the product undergoing challenge testing, and, if possible, the spores should be heat-shocked in the food to be studied.

Quantitative counts on the challenge suspensions may be conducted to aid in calculating the dilutions necessary to achieve the target inoculum in the challenge product. Appropriate procedures and containment facilities should be used when carrying out challenge tests with certain pathogens.TheMethod of inoculation is another extremelyimportant consideration when conducting amicrobiological challenge study.

Every effort must be made not to change the critical parameters of the product formulation undergoing challenge. There are a variety of inoculation methods that can be used depending upon the type of product being challenged. In aqueous liquid matrices such as sauces and gravies with high a_w (> 0.96), the challenge inoculum may be directly inoculated into the product with mixing, using a minimal amount of sterile water or buffer as a carrier.

Use of a diluent adjusted to the approximate a_w of the product using the humectant present in the food minimizes the potential for erroneous results in intermediate a_w foods. In studies where moisture level is one of the experimental variables, the inoculum may be suspended in the water or liquid used to adjust the moisture level of the formulation. For batch type inoculations, the inoculum may be added directly to the product in a mixing bowl or container. For individual package or pouch type applications, the inoculum may be aseptically injected using a sterile syringe through the package wall containing a rubber septum.

In solid matrices with a_w > 0.96, such as cooked pasta or meat surfaces, an alternative to the syringe method may be the use of an atomizer. An atomizer sprays the inoculum, which is suspended in sterile water or buffer, into the ground product or onto the surface of product. Spraying should be done in a containment hood or using other protective devices to avoid worker safety issues related to creation of pathogenic.

In all these applications, the smallest amount of water or buffer practical for suspension of the inoculum should be used. Inoculum may also be transferred using a velvet pad, paint pad, or similar fibrous cloth provided the method is calibrated and reproducible levels of inoculum can be delivered with minimum moisture transfer. Preliminary analyses should be done to ensure that the a_w or moisture level of the formulation is not changed after inoculation.

Products or components with a_w <0.92 may be inoculated using the atomizer method with a minimal volume of carrier water or buffer. Again, the product should always be checked to ensure that the final product a_w or moisture level has not been changed.

A short, post-inoculation drying period for some products may be needed prior to final packaging. Alternatively, they may be inoculated with challenge organisms that have been suspended in carrier water or buffer that has been added to sterile sand, flour, or a powdered form of the product (for example, dried pasta), and allowed to dry. Lyophilized culture may also be used for some applications. Inoculum viability and population levels should be determined in advance of the study. The dried inoculum preparation should be added aseptically to the test product and shaken or agitated thoroughly for even distribution of the inoculum.

Enough product should be inoculated so that a minimum of three replicates per sampling time is available throughout the challenge study. In some cases, such as in certain revalidation studies and for uninoculated control samples, fewer replicates may be used.

DURATION OF THE STUDY

It is Prudent to conduct the microbiological challenge study over, at least, the desired shelf life of the product. It is even more desirable to challenge the product for its entire desired shelf life plus a margin beyond the desired shelf life because it is important to determine what would happen if users would hold and consume the product beyond its intended shelf life. Some regulatory agencies require a minimum of data on shelf life plus at least one-third of the intended shelf life.

Another consideration impacting the duration of the challenge study is the temperature of product storage. Refrigerated products may be challenged for their entire shelf life under the target storage temperature, but under abuse temperatures they are typically held for shorter time.

In certain foodservice venues, it may be convenient for the food establishment to hold specific refrigerated products at room temperature for short periods of time. However, pathogens may be present on the cheese slices due to cross-contamination through handling in the restaurant, and therefore challenge testing will be needed to provide evidence that this practice is safe.

If the restaurant would like to hold the cheese slices at room temperature for an 8 h shift, the duration of the challenge study should be at least 12 h. This challenge study is performed to ensure that the rapid growth of pathogens does not occur if the cheese slices are cross-contaminated in the restaurant through handling.

It is also desirable to test the product over and significantly beyond its entire shelf life because sublethal injury may occur in some products. This can lead to a long lag period, where it may not be possible to culture the inoculum, but over time, a small number of the injured cells recover and grow in the product. This rebound, or "Phoenix" phenomenon, has been observed in a number of products (Jay). If the product is not tested for at least its entire shelf life, it is possible to miss the recovery and subsequent growth of the challenge organism late in its shelf life.

The Frequency of testing is governed by the duration of the microbiological challenge study. It is desirable to have a minimum of 5–7 data points over the shelf life in order to have a good indication of the inoculum behaviour. Typically, if the shelf life is measured in days, the frequency of testing should be at least daily, if not multiple times per day. If the shelf life is measured in weeks or months, the test frequency is typically no less than once per week.

All studies should start with "zero time" testing, that is, analysis of the product right after inoculation. For some types of products, it may be desirable to also allow an equilibration period for the inoculum to adapt to the product before testing. It may be desirable to test more frequently (for example, daily or multiple times per day) early in the challenge study (that is, for the first few days or week), and then reduce the frequency of testing to longer intervals.

FORMULATION FACTORS AND STORAGE CONDITIONS

When evaluating a formulation, it is important to understand the range of key factors that control its microbiological stability. Intrinsic factors such as pH, aw, or preservative level may be key to preventing the growth of pathogens or to preventing spoilage that would influence the safety of the product during its intended shelf life.

It is, therefore, important to test each key variable singly and/or in combination in the formulation under worst-case conditions. For example, if the target pH is 4.8 + 0.2 and the process capability is within that tolerance range, it is important to challenge the product on the high side of that range (that is, pH 5.0). Similarly, if sorbic acid is used at a level of 0.15 + 0.05 per cent, the product should be challenged at the low concentration of 0.10 per cent. This is recommended to ensure that the challenge study covers the process capability range for each critical factor in the formulation. Relevant intrinsic properties such as pH, aw, and salt level should be documented for each study for future comparison and reference.

Test samples should ideally be stored in the same packaging as intended for the commercial marketplace. If the commercial product is vacuum or MAP packaged, then the samples used in the microbiological challenge study should be packaged under the same conditions using the same packaging film.

The storage temperature used in the microbiological challenge study should include the typical temperature range at which the product is to be held and distributed. A refrigerated product that may be subject to temperature abuse should be challenged under representative abuse temperatures. Products that may encounter high humidity environments should also be challenged under those conditions (Notermans and others).

Some challenge studies may incorporate temperature cycling into their protocol. For example, the manufacturer may distribute a refrigerated product under well-controlled conditions for a portion of its shelf life, after which the product may be subjected to elevated temperatures immediately prior to and during use.

SAMPLE ANALYSIS

Typically, in a Microbiological challenge study, the levels of live challenge microorganisms are enumerated at each sampling point. Usually, it is desirable to have at least duplicate and, preferably, triplicate samples for analysis at each time point. In cases where higher levels of certainty are needed, a larger number of replicates should be used or the study should be replicated. The selection of enumeration media and method (for example, direct plating versus Most Probable Number) is dependent on the type of pathogens or surrogates used in the study.

If the product does not have a substantial background *microflora*, non-selective media for direct enumeration may be used. In cases where toxin-producing organisms are used (for example, *Staphylococcus aureus* or *C. botulinum*), appropriate toxin testing should be performed at each time point using the most current validated method. Toxin levels may not always be tested at each time point in the study, but should be done at frequent enough intervals throughout the desired shelf life of the product to determine if that shelf life is acceptable. Where appropriate, resucitation methods may be used to avoid erroneous result.

It is Prudent to analyse the product, including uninoculated control samples, at each or selected sampling points in the study to see how the background microflora is behaving over product shelf life. For example, if a product has a high background microflora, it may suppress the growth of the challenge inoculum. In some cases, this is useful and desirable because the product spoils before pathogens can grow. In other situations, the background microorganisms may not be universally present, leading to a potentially false sense of security. Also, under some circumstances, the background microorganisms can change the formulation parameters in the product to favour or inhibit growth of the inoculum over time (for example, molds can raise product pH; lactobacilli can decrease product pH).

It is also important to track pertinent physico-chemical parameters of the product over shelf life to see how they might change and influence the behaviour of the pathogen. Understanding how factors such as aw, moisture, salt level, pH, MAP gas concentrations, preservative levels, and other variables behave over product shelf life is key to understanding the microbiological stability of the product.

DATA INTERPRETATION

Once the microbiological challenge study is completed, the data should be analysed to see how the pathogens behaved over time.

Trend analysis and appropriate graphical plotting (that is, semi-log plots) of the data will show whether the challenge organisms died, remained stable, or increased in numbers over time. In the case of toxin-producing pathogens, no toxin should be detected over the designated challenge period. Combining the quantitative inoculum data for each time point with data on the background microflora and the relevant physico-chemical parameters gives a powerful and broad representation of the microbiological stability of the formulation under evaluation. Based on these data, a reasonable shelf life can be established or adjustments can be made to the formulation so that it is less susceptible to pathogen growth.

When using microbiological challenge testing, as part of a process validation protocol, analysis of the data will show whether the process is capable of delivering the required level of lethality (that is, conforms with the pre-determined performance standard). Based on this information, adjustments can be made to the process, if necessary, in order to meet the lethality requirements.

The Data from microbiological challenge testing can be used in developing predictive microbiological models or in validating existing ones. Predictive models are computer-based programmes that simulate or predict how specific microorganisms will behave in a formulation under specific conditions (for example, pH, a_w, moisture, salt, and preservatives). Microbiological challenge tests are used both to generate these types of empirical models and to validate their applicability.

Overall, well-designed challenge studies can provide critical information on the microbiological safety and stability of a food formulation. They are also invaluable in validating key lethality or microbiological control points in a process. Challenge studies can be an invaluable aid in determining if a food product requires temperature control throughout its shelf life or if it can tolerate storage at room temperature for a portion or all of its shelf life.

PASS/FAIL CRITERIA

Selection of microorganisms to use in challenge testing and/or modeling depends on the knowledge gained through commercial experience and/or on epidemiological data that indicate that the food under consideration or similar foods may be hazardous due to pathogen growth. In addition, the intrinsic properties (for example, pH, water activity, and preservatives) and extrinsic properties (for example, atmosphere, temperature, and processing) should be considered.

The significance of a population increase varies with the hazard characterization of each microorganism. For example, the growth of infectious pathogens should always be controlled, whereas most toxin production requires substantial growth before a hazard exists. In this case, growth of the toxigenic organism alone does not result in a health hazard, but toxin production will.

The following list identifies microorganisms that can be used in a microbiological challenge study along with the panel's recommendations and rationale for selection and assessment of tolerable growth. Toxigenic molds such as *Aspergillus, Penicillium*, and *Fusarium spp*. Challenge studies related to the need for time/temperature control for safety are not recommended because mold provides a visual clue to prevent consumption of the spoiled product.

Bacillus Cereus: The absence of toxin formation is the preferred criterion. However, since toxin measurement is difficult, a 3 log increase over inoculum levels would indicate the need for time/temperature control.

This growth limit determination is based on the following:

- Typical initial levels of B. cereus are low; therefore, 1000 CFU/g would be a conservative initial level based on the literature.
- Populations of > 106 CFU/g are needed to produce toxin at levels hazardous to health (FDA 2001).
- The emetic toxin of B. cereus is heat stable; therefore, no reduction is likely.

Other considerations:

- *Bacillus cereus* spores are relatively heat sensitive.
- Baking is likely to destroy low levels found in flour used in bread products (Kaur); however, unusual high heat resistance has been reported in pumpkin pie (Wyatt and Guy 1981).
- The potential for survival of B. cereus should be evaluated for specific products.
- Rice and potatoes have a history of association with *B. cereus* foodborne illness and therefore products containing rice or potatoes should be evaluated for time/temperature control requirements.

Campylobacter spp: No challenge testing is recommended because other organisms such as Salmonella have similar routes of contamination, are less fastidious, and are easier to culture.

Furthermore, its minimal growth temperature and water activity of 32°C (90 °F) and 0.98, respectively, make Campylobacter spp. it an unlikely candidate for challenge studies.

Clostridium botulinum: The absence of toxin formation based on current methodology is the recommended requirement. Other considerations: *C. botulinum* is appropriate to consider for certain products, particularly those packaged under anaerobic and micro-aerophilic conditions such as MAP products; and those with a history of associated illness, such as products under oil and potatoes.

Clostridium perfringens: *A 3 log increase is recommended based on the following facts*:

- Although other products may contain surviving spores, *Clostridium perfringens* is relevant mainly to meat and poultry products, including sauces and gravies. Most products subject to the Food Code requirements will be either raw or freshly cooked.
- Vegetative cells of *C. perfringens* are easily destroyed by cooking meat and poultry products, and spore levels are typically low due to demanding sporulation requirements. An initial population of 100 CFU/g was considered to be a conservative worst case by the panel. A

population of >105 CFU/g is needed to result in illness; therefore, a 3 log increase would control the hazard.

Enterohemorrhagic E. Coli: If modeling programmes are used to predict the growth of the pathogen, time/temperature holding conditions should maintain enterohemorrhagic *E. Coli* in lag phase due to the infectious nature of the microorganism. However, if laboratory challenge studies are used, the inherent variability in quantitative methods necessitates the use of a progressive increase of < 1 log as indicative that growth is controlled.

Listeria Monocytogenes: Recent risk assessments (FDA/USDA 2001) indicate that low numbers of *L. Monocytogenes* present a low risk to public health. In recognition of this, some countries such as Canada and Germany have established a tolerance for low levels of this organism that will not support growth to high levels.

However, a tolerance for *L. Monocytogenes* has not been established in the United States. It is also recognized that products that support the growth of the microorganism present an increased risk. A *L. Monocytogenes* level of 100 CFU/g at the time of consumption may provide an acceptable level of consumer protection (Ross and others). However, data are insufficient to determine general worst-case initial levels. Overall, the panel concluded that a 1 log increase was an appropriate level of control for *L. monocytogenes*. This level accounts for variability in enumeration techniques and represents a view that growth of this organism to high levels represents a risk to public health that must be controlled.

Salmonella spp: Appropriately validated pathogen modeling programmes for growth can be used to verify that*Salmonella* spp. is maintained in the lag phase. Otherwise, population growth should be limited to < 1 log, following the same rationale as for enterohemorrhagic *E. Coli*.

Shigella spp: No challenge studies are recommended for Shigella spp. because it has the same potential source as *Salmonella spp*. and has more fastidious growth and survival requirements.

Staphylococcus Aureus: No detectable toxin should be formed under the time/temperature studies evaluated. As with *C. Botulinum*, current methodology should be used for toxin detection and specific toxin levels should be determined. In lieu of testing for toxin, limiting growth to <3 logs may be used. This limiting growth level is based on an initial population of 1000 CFU/g, and a minimum of 10^6 CFU/g to produce toxin.

Other Considerations: *Staphylococcus aureus* is appropriate to study in foods that receive extensive handling because of the human source of the microorganism. *Staphylococcus aureus* does not compete well with other microorganisms; therefore, it is not appropriate to consider in foods with high levels of other organisms, such as raw vegetables or properly fermented products.

Vibrio spp: Appropriately validated pathogen modeling programmes for growth can be used to verify that *Vibrio*spp. is maintained in the lag phase. Otherwise, population growth should be limited to < 1 log, following the same rationale as for *E. Coli*.

Vibrio Parahaemolyticus: Can be used as a surrogate for other *Vibrio spp*. *Vibrio parahaemolyticus* studies are only appropriate for marine foods. It should also be noted that most fish are highly perishable and therefore will be temperature controlled for spoilage reasons.

Yersinia Enterocolitica: Challenge studies are not recommended as *Salmonella* spp. and *Yersinia enterocolitica* have similar sources and *Salmonellae* are easier to culture.

11

Allosteric Enzymes

INTRODUCTION

The extraordinary ability of an enzyme to catalyze only one particular reaction is a quality known as specificity. Specificity means an enzyme acts only on a specific substance, its substrate, invariably transforming it into a specific product. That is, an enzyme binds only certain compounds, and then, only a specific reaction ensues. Some enzymes show absolute specificity, catalyzing the transformation of only one specific substrate to yield a unique product. Other enzymes carry out a particular reaction but act on a class of compounds. For example, hexokinase (ATP: hexose-6-phosphotransferase) will carry out the ATP-dependent phosphorylation of a number of hexoses at the 6-position, including glucose.

SPECIFICITY

An enzyme molecule is typically orders of magnitude larger than its substrate. Its active site comprises only a small portion of the overall enzyme structure. The active site is part of the conformation of the enzyme molecule arranged to create a special pocket or cleft whose three-dimensional structure is complementary to the structure of the substrate. The enzyme and the substrate molecules "recognize" each other through this structural complementarity.

The substrate binds to the enzyme through relatively weak forces — H bonds, ionic bonds (salt bridges), and van der Waals interactions between sterically complementary clusters of atoms. Specificity studies on enzymes entail an examination of the rates of the enzymatic reaction obtained with various structural analogs of the substrate.

THE "LOCK AND KEY" HYPOTHESIS

Pioneering enzyme specificity studies at the turn of the century by the great organic chemist Emil Fischer led to the notion of an enzyme resembling a "lock" and its particular substrate the "key." This analogy captures the essence of the specificity that exists between an enzyme and its substrate, but enzymes are not rigid templates like locks.

THE "INDUCED FIT" HYPOTHESIS

Enzymes are highly flexible, conformationally dynamic molecules, and many of their remarkable properties, including substrate binding and catalysis, are due to their structural pliancy. Realization of the conformational flexibility of proteins led Daniel Koshland to hypothesize that the binding of a substrate (S) by an enzyme is an interactive process. That is, the shape of the enzyme's active site is actually modified upon binding S, in a process of dynamic recognition between enzyme and substrate aptly called induced fit. In essence, substrate binding alters the conformation of the protein, so that the protein and the substrate "fit" each other more precisely. The process is truly interactive in that the conformation of the substrate also changes as it adapts to the conformation of the enzyme.

This idea also helps to explain some of the mystery surrounding the enormous catalytic power of enzymes: In enzyme catalysis, precise orientation of catalytic residues comprising the active site is necessary for the reaction to occur; substrate binding induces this precise orientation by the changes it causes in the protein's conformation.

"INDUCED FIT" AND THE TRANSITION-STATE INTERMEDIATE

The catalytically active enzyme:substrate complex is an interactive structure in which the enzyme causes the substrate to adopt a form that mimics the transition-state intermediate of the reaction. Thus, a poor substrate would be one that was less effective in directing the formation of an optimally active enzyme:transition -state intermediate conformation. This active conformation of the enzyme molecule is thought to be relatively unstable in the absence of substrate, and free enzyme thus reverts to a conformationally different state.

CONTROLS OVER ENZYMATIC ACTIVITY

The activity displayed by enzymes is affected by a variety of factors, some of which are essential to the harmony of metabolism.

1. The enzymatic rate, $v = d[P]/dt$, "slows down" as product accumulates and equilibrium is approached. The apparent decrease in rate is due to the conversion of P to S by the reverse reaction as [P] rises. Once $[P]/[S] = K_{eq}$, no further reaction is apparent. Keq defines thermodynamic equilibrium. Enzymes have no influence on the thermodynamics of a reaction. Also, product inhibition can be a kinetically valid phenomenon: Some enzymes are actually inhibited by the products of their action.
2. The availability of substrates and cofactors will determine the

enzymatic reaction rate. In general, enzymes have evolved such that their Km values approximate the prevailing in vivo concentration of their substrates. (It is also true that the concentration of some enzymes in cells is within an order of magnitude or so of the concentrations of their substrates.)

3. There are genetic controls over the amounts of enzyme synthesized (or degraded) by cells. If the gene encoding a particular enzyme protein is turned on or off, changes in the amount of enzyme activity soon follow. Induction, which is the activation of enzyme synthesis, and repression, which is the shutdown of enzyme synthesis, are important mechanisms for the regulation of metabolism. By controlling the amount of an enzyme that is present at any moment, cells can either activate or terminate various metabolic routes. Genetic controls over enzyme levels have a response time ranging from minutes in rapidly dividing bacteria to hours (or longer) in higher eukaryotes.
4. Enzymes can be regulated by covalent modification, the reversible covalent attachment of a chemical group. For example, a fully active enzyme can be converted into an inactive form simply by the covalent attachment of a functional group, such as a phosphoryl moiety. Alternatively, some enzymes exist in an inactive state unless specifically converted into the active form through covalent addition of a functional group. Covalent modification reactions are catalyzed by special converter enzymes, which are themselves subject to metabolic regulation. Although covalent modification represents a stable alteration of the enzyme, a different converter enzyme operates to remove the modification, so that when the conditions that favored modification of the enzyme are no longer present, the process can be reversed, restoring the enzyme to its unmodified state. Many examples of covalent modification at important metabolic junctions will be encountered in our discussions of metabolic pathways. Because covalent modification events are enzyme-catalyzed, they occur very quickly, with response times of seconds or even less for significant changes in metabolic activity. The 1992 Nobel Prize in physiology or medicine was awarded to Edmond Fischer and Edwin Krebs for their pioneering studies of reversible protein phosphorylation as an important means of cellular regulation.

Enzymatic activity can also be activated or inhibited through non-covalent interaction of the enzyme with small molecules (metabolites) other than the substrate. This form of control is termed allosteric regulation, because the activator or inhibitor binds to the enzyme at a site other than (allo means "other") the active site. Further, such allosteric regulators, or effector molecules, are often quite different sterically from the substrate. Because this form of regulation results simply from reversible binding of regulatory ligands to the enzyme, the cellular response time can be virtually instantaneous.

Specialized controls: Enzyme regulation is an important matter to cells, and evolution has provided a variety of additional options, including zymogens, isozymes, and modulator proteins.

ZYMOGENS

Most proteins become fully active as their synthesis is completed and they spontaneously fold into their native, three-dimensional conformations. Some proteins, however, are synthesized as inactive precursors, called zymogens or proenzymes, that only acquire full activity upon specific proteolytic cleavage of one or several of their peptide bonds. Unlike allosteric regulation or covalent modification, zymogen activation by specific proteolysis is an irreversible process. Activation of enzymes and other physiologically important proteins by specific proteolysis is a strategy frequently exploited by biological systems to switch on processes at the appropriate time and place, as the following examples illustrate.

Insulin

Some protein hormones are synthesized in the form of inactive precursor molecules, from which the active hormone is derived by proteolysis. For instance, insulin, an important metabolic regulator, is generated by proteolytic excision of a specific peptide from proinsulin.

PROTEOLYTIC ENZYMES OF THE DIGESTIVE TRACT.

Enzymes of the digestive tract that serve to hydrolyze dietary proteins are synthesized in the stomach and pancreas as zymogens (Table).

Table. Pancreatic and Gastric Zymogens

Origin	Zymogen	Active Protease
Pancreas	Trypsinogen	Trypsin
Pancreas	Chymotrypsinogen	Chymotrypsin
Pancreas	Procarboxypeptidase	Carboxypeptidase
Pancreas	Proelastase	Elastase
Stomach	Pepsinogen	Pepsin

Only upon proteolytic activation are these enzymes able to form a catalytically active substrate-binding site. The activation of chymotrypsinogen is an interesting example. Chymotrypsino-gen is a 245-residue polypeptide chain cross-linked by five disulfide bonds. Chymotrypsinogen is converted to an enzymatically active form called ?-chymotrypsin when trypsin cleaves the peptide bond joining Arg15 and Ile16. The enzymatically active ?-chymotrypsin acts upon other ?-chymotrypsin molecules, excising two dipeptides, Ser14-Arg15 and Thr147-Asn148. The end product of this processing pathway is the mature protease a -chymotrypsin, in which the three peptide chains, A (residues 1 through 13), B (residues 16 through 146), and C (residues 149 through 245), remain together because they are linked by two disulfide bonds, one from A to B, and one from B to C. It is interesting to note that the transformation of inactive chymotrypsinogen to active ?-chymotrypsin requires the cleavage of just one particular peptide bond.

BLOOD CLOTTING.

The formation of blood clots is the result of a series of zymogen activations. The amplification achieved by this cascade of enzymatic activations allows blood clotting to occur rapidly in response to injury. Seven of the clotting factors in their active form are serine proteases: kallikrein, XIIa, XIa, IXa, VIIa, Xa, andthrombin.

Two routes to blood clot formation exist. The intrinsic pathway is instigated when the blood comes into physical contact with abnormal surfaces caused by injury; the extrinsic pathway is initiated by factors released from injured tissues. The pathways merge at Factor X and culminate in clot formation. Thrombin excises peptides rich in negative charge from fibrinogen, converting it to fibrin, a molecule with a different surface charge distribution. Fibrin readily aggregates into ordered fibrous arrays that are subsequently stabilized by covalent cross-links. Thrombin specifically cleaves Arg-Gly peptide bonds and is homologous to trypsin, which is also a serine protease (recall that trypsin acts only at Arg and Lys residues).

Protein Kinases

Protein kinases are converter enzymes that catalyze the ATP-dependent phosphorylation of serine, threonine, and/or tyrosine hydroxyl groups in target proteins (table). Phosphorylation introduces a bulky group bearing two negative charges, causing conformational changes that alter the target protein's function. (Unlike a phosphoryl group, no amino acid side chain can provide two negative charges.)

Protein kinases represent a protein superfamily whose members are widely diverse in terms of size, subunit structure, and subcellular localization. Nevertheless, all share a common catalytic mechanism based on a conserved catalytic core/kinase domain of approximately 260 amino acid residues. Protein kinases are classified as Ser/Thr- and/or Tyr-specific and are subclassified in terms of the allosteric activators they require and the consensus amino acid sequence within the target protein that is recognized by the kinase.

For example, cAMP-dependent protein kinase (PKA) phosphorylates proteins having Ser or Thr residues within an R(R/K)X (S*/T*) target consensus sequence (* denotes the residue that becomes phosphorylated). That is, PKA phosphorylates Ser or Thr residues that occur in an Arg-(Arg or Lys)-(any amino acid)-(Ser or Thr) sequence segment (table). Targeting of protein kinases to particular consensus sequence elements within proteins creates a means to regulate these kinases byintrasteric control.

Intrasteric control occurs when a regulatory subunit (or protein domain) has a pseudosubstrate sequence that mimics the target sequence but lacks a OH-bearing side chain at the right place. For example, the cAMP-binding regulatory subunits of PKA possess the pseudosubstrate sequence RRGA*I and this sequence binds to the active site of PKA catalytic subunits, blocking their activity. This pseudosubstrate sequence has an alanine residue where serine occurs in the PKA target sequence; Ala is sterically similar to serine but lacks a phosphorylatable OH-group. When these PKA regulatory subunits bind cAMP, they undergo a conformational change and dissociate from the catalytic (C) subunits, and the active site of PKA is free to bind and phosphorylate its targets.

The abundance of many protein kinases in cells is an indication of the great importance of protein phosphorylation in cellular regulation. Exactly 113 protein kinase genes have been recognized in yeast, and it is estimated that the human genome encodes more than 1000 different protein kinases. Tyrosine kinases(protein kinases that phosphorylate Tyr residues) occur only in multicellular organisms (yeast has no tyrosine kinases). Tyrosine kinases are components of signaling pathways involved in cell - cell communication.

ISOZYMES

A number of enzymes exist in more than one quaternary form, differing in their relative proportions of structurally equivalent but catalytically distinct polypeptide subunits. A classic example is mammalian lactate dehydrogenase (LDH), which exists as five different isozymes, depending on the tetrameric association of two different subunits, A and B: A4, A3B, A2B2, AB3, and B4. The kinetic properties of the various LDH isozymes differ in terms of their relative affinities for the various substrates and their sensitivity to inhibition by product. Different tissues express different isozyme forms, as appropriate to their particular metabolic needs. By regulating the relative amounts of A and B subunits they synthesize, the cells of various tissues control which isozymic forms are likely to assemble, and, thus, which kinetic parameters prevail.

MODULATOR PROTEINS

Modulator proteins are yet another way that cells mediate metabolic activity. Modulator proteins are proteins that bind to enzymes, and by binding, influence the activity of the enzyme. For example, some enzymes, such as cAMP-dependent protein kinase, exist as dimers of catalytic subunits and regulatory subunits. These regulatory subunits are modulator proteins that suppress the activity of the catalytic subunits. Dissociation of the regulatory subunits (modulator proteins) activates the catalytic subunits; reassociation once again suppresses activity.

Phosphoprotein phosphatase inhibitor-1 (PPI-1) is another example of a modulator protein. When PPI-1 is phosphorylated on one of its serine residues, it binds to phosphoprotein phosphatase, inhibiting its phosphatase activity. The result is an increased phosphorylation of the interconvertible enzyme targeted by the protein kinase/phosphoprotein phosphatase cycle. We will meet other important representatives of this class as the processes of metabolism unfold in subsequent chapters. For now, let us focus our attention on the fascinating kinetics of allosteric enzymes.

ALLOSTERIC REGULATION OF ENZYME ACTIVITY

Allosteric regulation acts to modulate enzymes situated at key steps in metabolic pathways. Consider as an illustration the following pathway, where A is the precursor for formation of an end product, F, in a sequence of five enzyme-catalyzed reactions:

enz 1 enz 2 enz 3 enz 4 enz 5

A ? B ? C ? D ? E ? F

In this scheme, F symbolizes an essential metabolite, such as an amino acid or a nucleotide. In such systems, F, the essential end product, inhibits enzyme 1, the first step in the pathway. Therefore, when sufficient F is synthesized, it blocks further synthesis of itself. This phenomenon is called feedback inhibition or feedback regulation.

GENERAL PROPERTIES OF REGULATORY ENZYMES

Enzymes such as enzyme 1, which are subject to feedback regulation, represent a distinct class of enzymes, the regulatory enzymes. As a class, these enzymes have certain exceptional properties:

1. Their kinetics do not obey the Michaelis - Menten equation. Their v versus [S] plots yield sigmoid- or S-shaped curves rather than rectangular hyperbolas. Such curves suggest a second-order (or higher) relationship between v and [S]; that is, v is proportional to [S]n, where n > 1. A qualitative description of the mechanism responsible for the S-shaped curves is that binding of one S to a protein molecule makes it easier for additional substrate molecules to bind to the same protein molecule. In the jargon of allostery, substrate binding is cooperative.

Inhibition of a regulatory enzyme by a feedback inhibitor does not conform to any normal inhibition pattern, and the feedback inhibitor F bears little structural similarity to A, the substrate for the regulatory enzyme. F apparently acts at a binding site distinct from the substrate-binding site. The term allosteric is apt, because F is sterically dissimilar and, moreover, acts at a site other than the site for S. Its effect is called allosteric inhibition.

Regulatory or allosteric enzymes like enzyme 1 are, in some instances, regulated by activation. That is, whereas some effector molecules such as F exert negative effects on enzyme activity, other effectors show stimulatory, or positive, influences on activity.

Allosteric enzymes have an oligomeric organization. They are composed of more than one polypeptide chain (subunit) and have more than one S-binding site per enzyme molecule.

The working hypothesis is that, by some means, interaction of an allosteric enzyme with effectors alters the distribution of conformational possibilities or subunit interactions available to the enzyme. That is, the regulatory effects exerted on the enzyme's activity are achieved by conformational changes occurring in the protein when effector metabolites bind.

MODELS FOR THE ALLOSTERIC BEHAVIOUR OF PROTEINS

THE SYMMETRY MODEL OF MONOD, WYMAN, AND CHANGEUX

In 1965, Jacques Monod, Jeffries Wyman, and Jean-Pierre Changeux proposed a theoretical model of allosteric transitions based on the observation that allosteric proteins are oligomers. They suggested that allosteric proteins can exist in (at least) two conformational states, designated R, signifying "relaxed," and T, or "taut," and that, in each protein molecule, all of the subunits have the same conformation (either R or T). That is, molecular symmetry is conserved. Molecules of mixed conformation (having subunits of both R and T states) are not allowed by this model.

In the absence of ligand, the two states of the allosteric protein are in equilibrium:

R0 ? T0

(Note that the subscript "0" signifies "in the absence of ligand.") The equilibrium constant is termed L: L = T0/R0. L is assumed to be large; that is, the amount of the protein in the T conformational state is much greater than the amount in the R conformation. Let us suppose that L = 104.

The affinities of the two states for substrate, S, are characterized by the respective dissociation constants, KR and KT. The model supposes that KT >> KR. That is, the affinity of R0 for S is much greater than the affinity of T0 for S. Let us choose the extreme where KR/KT = 0 (that is, KT is infinitely greater than KR). In effect, we are picking conditions where S binds only to R. (If KT is infinite, T does not bind S.)

Given these parameters, consider what happens when S is added to a solution of the allosteric protein at conformational equilibrium. Although the relative [R0] concentration is small, S will bind "only" to R0, forming R1. This depletes the concentration of R0, perturbing the T0/R0 equilibrium. To restore equilibrium, molecules in the T0 conformation undergo a transition to R0. This shift renders more R0 available to bind S, yielding R1, diminishing [R0], perturbing the T0/R0 equilibrium, and so on. Thus, these linked equilibria are such that S-binding by the R0 state of the allosteric protein perturbs the T0/R0 equilibrium with the result that S-binding drives the conformational transition, T0 ? R0.

Monod - Wyman - Changeux (MWC) model for allosteric transitions. Consider a dimeric protein that can exist in either of two conformational states, R or T. Each subunit in the dimer has a binding site for substrate S and an allosteric effector site, F. The promoters are symmetrically related to one another in the protein, and symmetry is conserved regardless of the conformational state of the protein. The different states of the protein, with or without bound ligand, are linked to one another through the various equilibria. Thus, the relative population of protein molecules in the R or T state is a function of these equilibria and the concentration of the various ligands, substrate (S), and effectors (which bind at FR or FT). As [S] is increased, the T/R equilibrium shifts in favour of an increased proportion of R-conformers in the total population (that is, more protein molecules in the R conformational state).

In just this simple system, cooperativity is achieved because each subunit has a binding site for S, and thus, each protein molecule has more than one binding site for S. Therefore, the increase in the population of R conformers gives a progressive increase in the number of sites available for S. The extent of cooperativity depends on the relative T0/R0 ratio and the relative affinities of R and T for S. If L is large (that is, the equilibrium lies strongly in favour of T0) and if KT >> KR, as in the example we have chosen, cooperativity is great. Ligands such as S here that bind in a cooperative manner, so that binding of one equivalent enhances the binding of additional equivalents of S to the same protein molecule, are termed positive homotropic effectors. (The prefix "homo" indicates that the ligand influences the binding of like molecules.)

The Monod - Wyman - Changeux model. Graphs of allosteric effects for a tetramer (n = 4) in terms of Y, the saturation function, versus [S]. Y is defined as [ligand-binding sites that are occupied by ligand]/[total ligand-binding sites]. (a) A plot of Y as a function of [S], at various L values. (b) Y as a function of [S], at different c, where c = KR/KT. (When c = 0, KT is infinite.)

HETEROTROPIC EFFECTORS

This simple system also provides an explanation for the more complex substrate-binding responses to positive and negative effectors. Effectors that influence the binding of something other than themselves are termed heterotropic effectors. For example, effectors that promote S binding are termed positive heterotropic effectors or allosteric activators. Effectors that diminish

S binding are negative heterotropic effectors or allosteric inhibitors. Feedback inhibitors fit this class. Consider a protein composed of two subunits, each of which has two binding sites: one for the substrate, S, and one to which allosteric effectors bind, the allosteric site. Assume that S binds preferentially ("only") to the R conformer; further assume that the positive heterotropic effector, A, binds to the allosteric site only when the protein is in the R conformation, and thenegative allosteric effector, I, binds at the allosteric site only if the protein is in the T conformation. Thus, with respect to binding at the allosteric site, A and I are competitive with each other.

POSITIVE EFFECTORS

If A binds to R0, forming the new species R1(A), the relative concentration of R0 is decreased and the T0/R0 equilibrium is perturbed. As a consequence, a relative T0 ? R0 shift occurs in order to restore equilibrium. The net effect is an increase in the number of R conformers in the presence of A, meaning that more binding sites for S are available. For this reason, A leads to a decrease in the cooperativity of the substrate saturation curve, as seen by a shift of this curve to the left. Effectively, the presence of A lowers the apparent value of L.

Heterotropic allosteric effects: A and I binding to R and T, respectively. The linked equilibria lead to changes in the relative amounts of R and T and, therefore, shifts in the substrate saturation curve. This behaviour, depicted by the graph, defines an allosteric "K" system. The parameters of such a system are: (1) S and A (or I) have different affinities for R and T and (2) A (or I) modifies the apparent K0.5 for S by shifting the relative R versus T population.

NEGATIVE EFFECTORS

The converse situation applies in the presence of I, which binds "only" to T. I-binding will lead to an increase in the population of T conformers, at the expense of R0 (Figure). The decline in [R0] means that it is less likely for S (or A) to bind. Consequently, the presence of I increases the cooperativity (that is, the sigmoidicity) of the substrate saturation curve, as evidenced by the shift of this curve to the right (Figure). The presence of I raises the apparent value of L.

K SYSTEMS AND V SYSTEMS

The allosteric model just presented is called a K system because the concentration of substrate giving half-maximal velocity, defined as K0.5, changes in response to effectors. Note that Vmax is constant in this system.

An allosteric situation where K0.5 is constant but the apparent Vmax changes in response to effectors is termed a V system. In a V system, all v versus S plots are hyperbolic rather than sigmoid. The positive heterotropic effector A activates by raising Vmax, whereas I, the negative heterotropic effector, decreases it. Note that neither A nor I affects K0.5. This situation arises if R and T have the same affinity for the substrate, S, but differ in their catalytic ability and their affinities for A and I. A and I thus can shift the relative T/R distribution.

Acetyl-coenzyme A carboxylase, the enzyme catalyzing the committed step in the fatty acid biosynthetic pathway, behaves as a V system in response to its allosteric activator, citrate. v versus [S] curves for an allosteric "V" system. The V system fits the model of Monod, Wyman, and Changeux, given the following conditions: (1) R and T have the same affinity for the substrate, S. (2) The effectors A and I have different affinities for R and T and thus can shift the relative T/R distribution. (That is, A and I change the apparent value of L.) Assume as before that A binds "only" to the R state and I binds "only" to the T state. (3) R and T differ in their catalytic ability. Assume that R is the enzymatically active form, whereas T is inactive. Because A perturbs the T/R equilibrium in favour of more R, A increases the apparent Vmax. I favours transition to the inactive T state.

K SYSTEMS AND V SYSTEMS FILL DIFFERENT BIOLOGICAL ROLES

The K and V systems have design features that mean they work best under different physiological situations. "K system" enzymes are adapted to conditions in which the prevailing substrate concentration is rate-limiting, as when [S] in vivo ? K0.5. On the other hand, when the physiological conditions are such that [S] is usually saturating for the regulatory enzyme of interest, the enzyme conforms to the "V system" mode in order to have an effective regulatory response.

GLYCOGEN PHOSPHORYLASE

THE GLYCOGEN PHOSPHORYLASE REACTION

The cleavage of glucose units from the non-reducing ends of glycogen molecules is catalyzed by glycogen phosphorylase, an allosteric enzyme. The enzymatic reaction involves phosphorolysis of the bond between C-1 of the departing glucose unit and the glycosidic oxygen, to yield glucose-1-phosphate and a glycogen molecule that is shortened by one residue.

(Because the reaction involves attack by phosphate instead of H_2O, it is referred to as aphosphorolysis rather than a hydrolysis.) Phosphorolysis produces a phosphorylated sugar product, glucose-1-P, which is converted to the glycolytic substrate, glucose-6-P, by phosphoglucomutase. In muscle, glucose-6-P proceeds into glycolysis, providing needed energy for muscle contraction. In liver, hydrolysis of glucose-6-P yields glucose, which is exported to other tissues via the circulatory system.

STRUCTURE OF GLYCOGEN PHOSPHORYLASE

Muscle glycogen phosphorylase is a dimer of two identical subunits (842 residues, 97.44 kD). Each subunit contains a pyridoxal phosphate cofactor, covalently linked as a Schiff base to Lys680. Each subunit contains an active site (at the center of the subunit) and an allosteric effector site near the subunit interface. In addition, a regulatory phosphorylation site is located at Ser14 on each subunit. A glycogen-binding site on each subunit facilitates prior association of glycogen phosphorylase with its substrate and also exerts regulatory control on the enzymatic reaction.

Each subunit contributes a tower helix to the subunit - subunit contact interface in glycogen phosphorylase. In the phosphorylase dimer, the tower helices extend from their respective subunits and pack against each other in an antiparallel manner.

REGULATION OF GLYCOGEN PHOSPHORYLASE BY ALLOSTERIC EFFECTORS

Muscle Glycogen Phosphorylase Shows Cooperativity in Substrate Binding

The binding of the substrate inorganic phosphate (Pi) to muscle glycogen phosphorylase is highly cooperative, which allows the enzyme activity to increase markedly over a rather narrow range of substrate concentration. Pi is a positive homotropic effectorwith regard to its interaction with glycogen phosphorylase. v versus S curves for glycogen phosphorylase. (a) The sigmoid response of glycogen phosphorylase to the concentration of the substrate phosphate (Pi) shows strong positive cooperativity. (b) ATP is a feedback inhibitor that affects the affinity of glycogen phosphorylase for its substrates but does not affect Vmax. (c) AMP is a positive heterotropic effector for glycogen phosphorylase. It binds at the same site as ATP. AMP and ATP are competitive. Like ATP, AMP affects the affinity of glycogen phosphorylase for its substrates, but does not affect Vmax.

ATP and Glucose-6-P Are Allosteric Inhibitors of Glycogen Phosphorylase

ATP can be viewed as the "end product" of glycogen phosphorylase action, in that the glucose-1-P liberated by glycogen phosphorylase is degraded in muscle via metabolic pathways whose purpose is energy (ATP) production. Glucose-1-P is readily converted into glucose-6-P to feed such pathways. (In the liver, glucose-1-P from glycogen is converted to glucose and released into the bloodstream to raise blood glucose levels.)

Thus, feedback inhibition of glycogen phosphorylase by ATP and glucose-6-P provides a very effective way to regulate glycogen breakdown. Both ATP and glucose-6-P act by decreasing the affinity of glycogen phosphorylase for its substrate Pi. Because the binding of ATP or glucose-6-P has a negative effect on substrate binding, these substances act as negative heterotropic effectors. The substrate saturation curve is displaced to the right in the presence of ATP or glucose-6-P, and a higher substrate concentration is needed to achieve half-maximal velocity (Vmax/2). When concentrations of ATP or glucose-6-P accumulate to high levels, glycogen phosphorylase is inhibited; when [ATP] and [glucose-6-P] are low, the activity of glycogen phosphorylase is regulated by availability of its substrate, Pi.

AMP Is an Allosteric Activator of Glycogen Phosphorylase

AMP also provides a regulatory signal to glycogen phosphorylase. It binds to the same site as ATP, but it stimulates glycogen phosphorylase rather than inhibiting it. AMP acts as a positive heterotropic effector, meaning that it enhances the binding of substrate to glycogen phosphorylase. Significant levels of AMP indicate that the energy status of the cell is low and that more energy (ATP) should be produced. Reciprocal changes in the cellular concentrations of ATP and AMP and their competition for binding to the same site (the allosteric site) on glycogen phosphorylase, with opposite effects, allow these two nucleotides to exert rapid and reversible control over glycogen phosphorylase activity. Such reciprocal regulation ensures that the production of energy (ATP) is commensurate with cellular needs.

To summarize, muscle glycogen phosphorylase is allosterically activated by AMP and inhibited by ATP and glucose-6-P; caffeine can also act as an allosteric inhibitor. When ATP and glucose-6-P are abundant, glycogen breakdown is inhibited. When cellular energy reserves are low (*i.e.*, high [AMP] and low [ATP] and [G-6-P]), glycogen catabolism is stimulated.

The mechanism of covalent modification and allosteric regulation of glycogen phosphorylase. The T states are blue and the R states blue-green.

Glycogen phosphorylase conforms to the Monod - Wyman - Changeux model of allosteric transitions, with the active form of the enzyme designated the R state and the inactive form denoted as the T state. Thus, AMP promotes the conversion to the active R state, whereas ATP, glucose-6-P, and caffeine favour conversion to the inactive T state.

X-ray diffraction studies of glycogen phosphorylase in the presence of allosteric effectors have revealed the molecular basis for the T ? R conversion. Although the structure of the central core of the phosphorylase subunits is identical in the T and R states, a significant change occurs at the subunit interface between the T and R states. This conformation change at the subunit interface is linked to a structural change at the active site that is important for catalysis. In the T state, the negatively charged carboxyl group of Asp283 faces the active site, so that binding of the anionic substrate phosphate is unfavorable. In the conversion to the R state, Asp283 is displaced from the active site and replaced by Arg569. The exchange of negatively charged aspartate for positively charged arginine at the active site provides a favorable binding site for phosphate. These allosteric controls serve as a mechanism for adjusting the activity of glycogen phosphorylase to meet normal metabolic demands. However, in crisis situations in which abundant energy (ATP) is needed immediately, these controls can be overridden by covalent modification of glycogen phosphorylase. Covalent modification through phosphorylation of Ser14 in glycogen phosphorylase converts the enzyme from a less active, allosterically regulated form (the b form) to a more active, allosterically unresponsive form (the aform). Covalent modification is like a "permanent" allosteric transition that is independent of [allosteric effector], such as AMP.

REGULATION OF GLYCOGEN PHOSPHORYLASE BY COVALENT MODIFICATION

As early as 1938, it was known that glycogen phosphorylase existed in two forms: the less active phosphorylase b and the more activephosphorylase a. In 1956, Edwin Krebs and Edmond Fischer reported that a "converting enzyme" could convert phosphorylase b to phosphorylase a. Three years later, Krebs and Fischer demonstrated that the conversion of phosphorylase b

Phosphorylation of Ser14 causes a dramatic conformation change in phosphorylase. Upon phosphorylation, the amino-terminal end of the protein (including residues 10 through 22) swings through an arc of 120°, moving into the subunit interface. This conformation change moves Ser14 by more than 3.6 nm.

Dephosphorylation of glycogen phosphorylase is carried out by phosphoprotein phosphatase 1. The action of phosphoprotein phosphatase 1 inactivates glycogen phosphorylase.

ENZYME CASCADES REGULATE GLYCOGEN PHOSPHORYLASE

The phosphorylation reaction that activates glycogen phosphorylase is mediated by an enzyme cascade. The first part of the cascade leads to hormonal stimulation (described in the next section) of adenylyl cyclase, a membrane-bound enzyme that converts ATP to adenosine-3',5' -cyclic monophosphate, denoted as cyclic AMP or simply cAMP. This regulatory molecule is found in all eukaryotic cells and acts as an intracellular messenger molecule, controlling a wide variety of processes. Cyclic AMP is known as a second messenger because it is the intracellular agent of a hormone (the "first messenger").

The adenylyl cyclase reaction yields 3',5' -cyclic AMP and pyrophosphate. The reaction is driven forward by subsequent hydrolysis of pyrophosphate by the enzyme inorganic pyrophosphatase.

The hormonal stimulation of adenylyl cyclase is effected by a transmembrane signaling pathway consisting of three components, all membrane-associated. Binding of hormone to the external surface of a hormone receptor causes a conformational change in this transmembrane protein, which in turn stimulates a GTP-binding protein (abbreviated G protein). G proteins are heterotrimeric proteins consisting of a - (45 - 47 kD), ?- (35 kD), and g - (7 - 9 kD) subunits. The a -subunit binds GDP or GTP and has an intrinsic, slow GTPase activity.

In the inactive state, the G a ? g complex has GDP at the nucleotide site. When a G protein is stimulated by a hormone-receptor complex, GDP dissociates and GTP binds to G a, causing it to dissociate from G? g and to associate with adenylyl cyclase. Binding of G a (GTP) activates adenylyl cyclase to form cAMP from ATP. However, the intrinsic GTPase activity of G aeventually hydrolyzes GTP to GDP, leading to dissociation of G a (GDP) from adenylyl cyclase and reassociation with G a ? g to form the inactive G a ? g complex. This cascade amplifies the hormonal signal because a single hormone-receptor complex can activate many G proteins before the hormone dissociates from the receptor, and because the G a -activated adenylyl cyclase can synthesize many cAMP molecules before bound GTP is hydrolyzed by G a. More than 100 different G protein - coupled receptors and at least 21 distinct G aproteins are known.

Hormone (H) binding to its receptor (R) creates a hormone;receptor complex (H:R) that catalyzes GDP-GTP exchange on the a -subunit of the heterotrimer G protein (G a ? g), replacing GDP with GTP. The G a -subunit with GTP bound dissociates from the ? g -subunits and binds to adenylyl cyclase (AC). AC becomes active upon association with G a:GTP and catalyzes the formation of cAMP from ATP. With time, the intrinsic GTPase activity of the G a -subunit hydrolyzes the bound GTP, forming GDP; this leads to dissociation of G a:GDP from AC, reassociation of G a with the ? g subunits, and cessation of AC activity. AC and the hormone receptor H are integral plasma membrane proteins; G a and G?g are membrane-anchored proteins.

Cyclic AMP is an essential activator of cAMP-dependent protein kinase (PKA). This enzyme is normally inactive because its two catalytic subunits (C) are strongly associated with a pair of regulatory subunits (R), which serve to block activity. Binding of cyclic AMP to the regulatory subunits induces a conformation change that causes the dissociation of the C monomers from the R dimer. The free C subunits are active and can phosphorylate other proteins. One of the many proteins phosphorylated by PKA is phosphorylase kinase. Phosphorylase kinase is inactive in the unphosphorylated state and active in the phosphorylated form. As its name implies, phosphorylase kinase functions to phosphorylate (and activate) glycogen phosphorylase. Thus, stimulation of adenylyl cyclase leads to activation of glycogen breakdown.

HEMOGLOBIN AND MYOGLOBIN

Ancient life forms evolved in the absence of oxygen and were capable only of anaerobic metabolism. As the earth's atmosphere changed over time, so too did living things. Indeed, the production of 02 by photosynthesis was a major factor in altering the atmosphere. Evolution to an oxygen-based metabolism was highly beneficial. Aerobic metabolism of sugars, for example, yields far more energy than corresponding anaerobic processes. Two important oxygen-binding proteins appeared in the course of evolution so that aerobic metabolic processes were no longer limited by the solubility of 02 in water. These proteins are represented in animals as hemoglobin (Hb) in blood and myoglobin (Mb) in muscle. Because hemoglobin and myoglobin are two of the most-studied proteins in nature, they have become paradigms of protein structure and function. Moreover, hemoglobin is a model for protein quaternary structure and allosteric function. The binding of 02 by hemoglobin, and its modulation by effectors such as protons, CO2, and 2,3 -bisphosphoglycerate, depend on interactions between subunits in the Hb tetramer. Subunit - subunit interactions in Hb reveal much about the functional significance of quaternary associations and allosteric regulation.

THE COMPARATIVE BIOCHEMISTRY OF MYOGLOBIN AND HEMOGLOBIN

A comparison of the properties of hemoglobin and myoglobin offers insights into allosteric phenomena, even though these proteins are notenzymes. Hemoglobin displays sigmoid-shaped 02-binding curves. The unusual shape of these curves was once a great enigma in biochemistry. Such curves closely resemble allosteric enzyme:substrate saturation graphs. In contrast, myoglobin's interaction with oxygen obeys classical Michaelis - Menten-type substrate saturation behaviour.

Before examining myoglobin and hemoglobin in detail, let us first encapsulate the lesson: Myoglobin is a compact globular protein composed of a single polypeptide chain 153 amino acids in length; its molecular mass is 17.2 kD. It contains heme, a porphyrin ring system complexing an iron ion, as its prosthetic group. Oxygen binds to Mb via its heme.

Hemoglobin (Hb) is also a compact globular protein, but Hb is a tetramer. It consists of four polypeptide chains, each of which is very similar structurally to the myoglobin polypeptide chain, and each bears a heme group. Thus, a hemoglobin molecule can bind four 02 molecules. In adult human Hb, there are two identical chains of 141 amino acids, the ?-chains, and two identical ?-chains, each of 146 residues.

The human Hb molecule is an?2?2-type tetramer of molecular mass 64.45 kD. The tetrameric nature of Hb is crucial to its biological function: When a molecule of O2binds to a heme in Hb, the heme Fe ion is drawn into the plane of the porphyrin ring. This slight movement sets off a chain of conformational events that are transmitted to adjacent subunits, dramatically enhancing the affinity of their heme groups for O2. That is, the binding of O2 to one heme of Hb makes it easier for the Hb molecule to bind additional equivalents of O2. Hemoglobin is a marvelously constructed molecular machine. Let us dissect its mechanism, beginning with its monomeric counterpart, the myoglobin molecule.

The myoglobin and hemoglobin molecules. Myoglobin (sperm whale): one polypeptide chain of 153 aa residues (mass 5 17.2 kD) has one heme (mass = 652 D) and binds one O2. Hemoglobin (human): four polypeptide chains, two of 141 aa residues (a) and two of 146 residues (?); mass = 64.45 kD. Each polypeptide has a heme; the Hb tetramer binds four O2. (Irving Geis)

MYOGLOBIN

Myoglobin is the oxygen-storage protein of muscle. The muscles of diving mammals such as seals and whales are especially rich in this protein, which serves as a store for O2 during the animal's prolonged periods underwater. Myoglobin is abundant in skeletal and cardiac muscle of non-diving animals as well. Myoglobin is the cause of the characteristic red colour of muscle.

THE MB POLYPEPTIDE CRADLES THE HEME GROUP

The myoglobin polypeptide chain is folded to form a cradle (4.4 x 4.4 x 2.5 nm) that nestles the heme prosthetic group. O2 binding depends on the heme's oxidation state. The iron ion in the heme of myoglobin is in the +2 oxidation state, that is, the ferrous form. This is the form that binds O2. Oxidation of the ferrous form to the +3 ferric form yieldsmetmyoglobin, which will not bind O2.

It is interesting to note that free heme in solution will readily interact with O2 also, but the oxygen quickly oxidizes the iron atom to the ferric state. Fe3+:protoporphyrin IX is referred to as hematin. Thus, the polypeptide of myoglobin may be viewed as serving three critical functions: it cradles the heme group, it protects the heme iron atom from oxidation, and it provides a pocket into which the O2 can fit.

Detailed structure of the myoglobin molecule. The myoglobin polypeptide chain consists of eight helical segments, designated by the letters A through H, counting from the N-terminus. These helices, ranging in length from 7 to 26 residues, are linked by short, unordered regions that are named for the helices they connect, as in the AB region or the EF region. The individual amino acids in the polypeptide are indicated according to their position within the various segments, as in His F8, the eighth residue in helix F, or Phe CD1, the first amino acid in the interhelical CD region. Occasionally, amino acids are specified in the conventional way, that is, by the relative position in the chain, as in Gly153. The heme group is cradled within the folded polypeptide chain. (Irving Geis)

O2 BINDING TO MB

Iron ions, whether ferrous or ferric, prefer to interact with six ligands, four of which share a common plane. The fifth and sixth ligands lie above and below this plane. In heme, four of the ligands are provided by the nitrogen atoms of the four pyrroles. A fifth ligand is donated by the imidazole side chain of amino acid residue His F8. When myoglobin binds O2 to become oxymyoglobin, the O2 molecule adds to the heme iron ion as the sixth ligand. O2 adds end on to the heme iron, but it is not oriented perpendicular to the plane of the heme. Rather, it is tilted about 60° with respect to the perpendicular.

The six liganding positions of an iron ion. Four ligands lie in the same plane; the remaining two are, respectively, above and below this plane. In myoglobin, His F8 is the fifth ligand; in oxymyoglobin, O2 becomes the sixth.

In deoxymyoglobin, the sixth ligand position is vacant, and in metmyoglobin, a water molecule fills the O2 site and becomes the sixth ligand for the ferric atom. On the oxygen-binding side of the heme lies another histidine residue, His E7. While its imidazole function lies too far away to interact with the Fe atom, it is close enough to contact the O2.

Therefore, the O2-binding site is a sterically hindered region. Biologically important properties stem from this hindrance. For example, the affinity of free heme in solution for carbon monoxide (CO) is 25,000 times greater than its affinity for O2. But CO only binds 250 times more tightly than O2 to the heme of myoglobin, because His E7 forces the CO molecule to tilt away from a preferred perpendicular alignment with the plane of the heme. This diminished affinity of myoglobin for CO guards against the possibility that traces of CO produced during metabolism might occupy all of the heme sites, effectively preventing O2 from binding. Nevertheless, CO is a potent poison and can cause death by asphyxiation.

Oxygen and carbon monoxide binding to the heme group of myoglobin.

O2 BINDING ALTERS MB CONFORMATION

What happens when the heme group of myoglobin binds oxygen? X-ray crystallography has revealed that a crucial change occurs in the position of the iron atom relative to the plane of the heme. In deoxymyoglobin, the ferrous ion has but five ligands, and it lies 0.055 nm above the plane of the heme, in the direction of His F8.

The iron:porphyrin complex is therefore dome-shaped. When O2 binds, the iron atom is pulled back towards the porphyrin plane and is now displaced from it by only 0.026 nm. The consequences of this small motion are trivial as far as the biological role of myoglobin is concerned. However, as we shall soon see, this slight movement profoundly affects the properties of hemoglobin. Its action on His F8 is magnified through changes in polypeptide conformation that alter subunit interactions in the Hb tetramer. These changes in subunit relationships are the fundamental cause of the allosteric properties of hemoglobin.

The displacement of the Fe ion of the heme of deoxymyoglobin from the plane of the porphyrin ring system by the pull of His F8. In oxymyoglobin, the bound O2 counteracts this effect.

THE PHYSIOLOGICAL SIGNIFICANCE OF COOPERATIVE

The relative oxygen affinities of hemoglobin and myoglobin reflect their respective physiological roles. Myoglobin, as an oxygen storage protein, has a greater affinity for O2 than hemoglobin at all oxygen pressures. Hemoglobin, as the oxygen carrier, becomes saturated with O2 in the lungs, where the partial pressure of O2 (pO2) is about 100 torr.1 In the capillaries of tissues, pO2 is typically 40 torr, and oxygen is released from Hb. In muscle, some of it can be bound by myoglobin, to be stored for use in times of severe oxygen deprivation, such as during strenuous exercise.

THE STRUCTURE OF THE HEMOGLOBIN MOLECULE

As noted, hemoglobin is an a 2 ?2 tetramer. Each of the four subunits has a conformation virtually identical to that of myoglobin. Two different types of subunits, a and ?, are necessary to achieve cooperative O2-binding by Hb. The ?-chain at 146 amino acid residues is shorter than the myoglobin chain (153 residues), mainly because its final helical segment (the H helix) is shorter. The a -chain (141 residues) also has a shortened H helix and lacks the D helix as well.

Conformational drawings of the a - and ?-chains of Hb and the myoglobin chain. (Irving Geis)

Max Perutz, who has devoted his life to elucidating the atomic structure of Hb, noted very early in his studies that the molecule was highly symmetrical. The actual arrangement of the four subunits with respect to one another is shown in Figure for horse methemoglobin.

All vertebrate hemoglobins show a three-dimensional structure essentially the same as this. The subunits pack in a tetrahedral array, creating a roughly spherical molecule 6.4 x 5.5 x 5.0 nm. The four heme groups, nestled within the easily recognizable cleft formed between the E and F helices of each polypeptide, are exposed at the surface of the molecule. The heme groups are quite far apart; 2.5 nm separates the closest iron ions, those of hemes a 1 and ?2, and those of hemes a 2 and ?1. The subunit interactions are mostly between dissimilar chains: each of the a -chains is in contact with both ?-chains, but there are few a-a or ?-? interactions.

The arrangement of subunits in horse methemoglobin, the first hemoglobin whose structure was determined by X-ray diffraction. The iron atoms on metHb are in the oxidized, ferric (Fe3+) state. (Irving Geis)

The Physiological Significance of the Hb:O2 Interaction

We can determine quantitatively the physiological significance of the sigmoid nature of the hemoglobin oxygen-binding curve, or, in other words, the biological importance of cooperativity. The coefficient n is the Hill coefficient, an index of the cooperativity (sigmoidicity) of the hemoglobin oxygen-binding curve. Taking p O2 in the lungs as 100 torr, P50 as 26 torr, and n as 2.8, Y, the fractional saturation of the hemoglobin heme groups with O2, is 0.98. If p O2 were to fall to 10 torr within the capillaries of an exercising muscle, Y would drop to 0.06. The oxygen delivered under these conditions would be proportional to the difference, Ylungs - Ymuscle, which is 0.92. That is, virtually all the oxygen carried by Hb would be released. Suppose instead that hemoglobin binding of O2 were not cooperative; in that case, the hemoglobin oxygen-binding curve would be hyperbolic, and n = 1.0. Then Y in the lungs would be 0.79 and Y in the capillaries, 0.28, and the difference in Y values would be 0.51. Thus, under these conditions, the cooperativity of oxygen binding by Hb means that 0.92/0.51 or 1.8 times as much O2can be delivered.

OXYGENATION MARKEDLY ALTERS THE QUATERNARY STRUCTURE OF HB

Crystals of deoxyhemoglobin shatter when exposed to O2. Further, X-ray crystallographic analysis reveals that oxy- and deoxyhemoglobin differ markedly in quaternary structure. In particular, specific a ?-subunit interactions change. The ab contacts are of two kinds. The a 1 ?1and a 2 ?2 contacts involve helices B, G, and H and the GH corner. These contacts are extensive and important to subunit packing; they remain unchanged when hemoglobin goes from its deoxy to its oxy form. The a 1 ?2 anda 2 ?1 contacts are called sliding contacts. They principally involve helices C and G and the FG corner.

Side view of one of the two a ? dimers in Hb, with packing contacts indicated in blue. The sliding contacts made with the other dimer are shown.

When hemoglobin undergoes a conformational change as a result of ligand binding to the heme, these contacts are altered. Hemoglobin, as a conformationally dynamic molecule, consists of two dimeric halves, an a1 ?1-subunit pair and an a 2 ?2-subunit pair. Each a ? dimer moves as a rigid body, and the two halves of the molecule slide past each other upon oxygenation of the heme. The two halves rotate some 15° about an imaginary pivot passing through the a ?-subunits; some atoms at the interface between a ? dimers are relocated by as much as 0.6 nm.

Movement of the Heme Iron by Less Than 0.04 nm Induces the Conformational Change in Hemoglobin

In deoxyhemoglobin, histidine F8 is liganded to the heme iron ion, but steric constraints force the Fe2+:His -N bond to be tilted about 8° from the perpendicular to the plane of the heme. Steric repulsion between histidine F8 and the nitrogen atoms of the porphyrin ring system, combined with electrostatic repulsions between the electrons of Fe2+ and the porphyrin ?-electrons, forces the iron atom to lie out of the porphyrin plane by about 0.06 nm. Changes in electronic and steric factors upon heme oxygenation allow the Fe2+ atom to move about 0.039 nm closer to the plane of the porphyrin, so now it is displaced only 0.021 nm above the plane. It is as if the 02 were drawing the heme Fe2+ into the porphyrin plane. This modest displacement of 0.039 nm seems a trivial distance, but its biological consequences are far-reaching. As the iron atom moves, it drags histidine F8 along with it, causing helix F, the EF corner, and the FG corner to follow. These shifts are transmitted to the subunit interfaces, where they trigger conformational readjustments that lead to the rupture of interchain salt links.

THE OXY AND DEOXY FORMS OF HEMOGLOBIN REPRESENT

Hemoglobin resists oxygenation because the deoxy form is stabilized by specific hydrogen bonds and salt bridges (ion-pair bonds). All of these interactions are broken in oxyhemoglobin, as the molecule stabilizes into a new conformation.

A crucial H bond in this transition involves a particular tyrosine residue. Both a - and ?-subunits have Tyr as the penultimate C-terminal residue (Tyr a 140 = Tyr HC2; Tyr ?145 = Tyr HC2, respectively2). The phenolic -OH groups of these Tyr residues form intrachain H bonds to the peptide C=O function contributed by Val FG5 in deoxyhemoglobin. (Val FG5 is a 93 and ?98, respectively.) The shift in helix F upon oxygenation leads to rupture of this Tyr HC2:Val FG5 hydrogen bond.

Further, eight salt bridges linking the polypeptide chains are broken as hemoglobin goes from the deoxy to the oxy form. Six of these salt links are between different subunits. Four of these six involve either carboxyl-terminal or amino-terminal amino acids in the chains; two are between the amino termini and the carboxyl termini of the a -chains, and two join the carboxyl termini of the ?-chains to the e -NH31 groups of the two Lys a 140 residues. The other two interchain electrostatic bonds link Arg and Asp residues in the two a -chains. In addition, ionic interactions between Asp ?94 and His ?146 form an intrachain salt bridge in each ?-subunit. In deoxyhemoglobin, with all of these interactions intact, the C-termini of the four subunits are restrained, and this conformational state is termed T, the tense or taut form. In oxyhemoglobin, these C-termini have almost complete freedom of rotation, and the molecule is now in its R, or relaxed, form.

Salt bridges between different subunits in hemoglobin. These non-covalent, electrostatic interactions are disrupted upon oxygenation. Arga141 and His?146 are the C-termini of the a - and ?-polypeptide chains. (a) The various intrachain and interchain salt links formed among the a - and ?-chains of deoxyhemoglobin. (b) A focus on those salt bridges and hydrogen bonds involving interactions between N-terminal and C-terminal residues in the a -chains. Note the Cl- ion, which bridges ionic interactions between the N-terminus of a 2 and the R group of Arga141. (c) A focus on the salt bridges and hydrogen bonds in which the residues located at the C-termini of ?-chains are involved. All of these links are abolished in the deoxy to oxy transition. (Irving Geis)

MODEL FOR THE ALLOSTERIC BEHAVIOUR OF HEMOGLOBIN

A model for the allosteric behaviour of hemoglobin is based on recent observations that oxygen is accessible only to the heme groups of the a -chains when hemoglobin is in the T conformational state. Perutz has pointed out that the heme environment of ?-chains in the T state is virtually inaccessible because of steric hindrance by amino acid residues in the E helix. This hindrance disappears when the hemoglobin molecule undergoes transition to the R conformational state.

Binding of O2 to the ?-chains is thus dependent on the T to R conformational shift, and this shift is triggered by the subtle changes that occur when O2 binds to the a -chain heme groups.

Changes in the Heme Iron upon O2 Binding

In deoxyhemoglobin, the six d electrons of the heme Fe21 exist as four unpaired electrons and one electron pair, and five ligands can be accommodated: the four N-atoms of the porphyrin ring system and histidine F8. In this electronic configuration, the iron atom is paramagnetic and in the high-spin state. When the heme binds O2as a sixth ligand, these electrons are rearranged into three e2 pairs and the iron changes to the low-spin state and is diamagnetic. This change in spin state allows the bond between the Fe21 ion and histidine F8 to become perpendicular to the heme plane and to shorten. In addition, interactions between the porphyrin N-atoms and the iron strengthen. Also, high-spin Fe21 has a greater atomic volume than low-spin Fe21 because its four unpaired e2 occupy four orbitals rather than two when the electrons are paired in low-spin Fe21.

So, low-spin iron is less sterically hindered and able to move nearer to the porphyrin plane.

H+ PROMOTES THE DISSOCIATION OF OXYGEN FROM HEMOGLOBIN

Protons, carbon dioxide, and chloride ions, as well as the metabolite 2,3 - bisphosphoglycerate (or BPG), all affect the binding of O2 by hemoglobin. Their effects have interesting ramifications, which we shall see as we discuss them in turn. Deoxyhemoglobin has a higher affinity for protons than oxyhemoglobin.

Thus, as the pH decreases, dissociation of O_2 from hemoglobin is enhanced. In simple symbolism, ignoring the stoichiometry of O_2 or H+ involved:

$$HbO_2 + H+ ? HbH+ + O_2$$

Expressed another way, H+ is an antagonist of oxygen binding by Hb, and the saturation curve of Hb for O_2 is displaced to the right as acidity increases. This phenomenon is called the Bohr effect, after its discoverer, the Danish physiologist Christian Bohr (the father of Niels Bohr, the atomic physicist). The effect has important physiological significance because actively metabolizing tissues produce acid, promoting O_2 release where it is most needed. About two protons are taken up by deoxyhemoglobin. The N-termini of the two a -chains and the His ?146 residues have been implicated as the major players in the Bohr effect. (The pKa of a free amino terminus in a protein is about 8.0, but the pKa of a protein histidine imidazole is around 6.5.) Neighbouring carboxylate groups of Asp ?94 residues help to stabilize the protonated state of the His ?146 imidazoles that occur in deoxyhemoglobin. However, when Hb binds O_2, changes in the conformation of ?-chains upon Hb oxygenation move the negative Asp function away, and dissociation of the imidazole protons is favored.

CO_2 ALSO PROMOTES THE DISSOCIATION OF O_2 FROM HEMOGLOBIN

Carbon dioxide has an effect on O2 binding by Hb that is similar to that of H+, partly because it produces H+ when it dissolves in the blood:

The enzyme carbonic anhydrase promotes the hydration of CO2. Many of the protons formed upon ionization of carbonic acid are picked up by Hb as O2 dissociates.

The bicarbonate ions are transported with the blood back to the lungs. When Hb becomes oxygenated again in the lungs, H1 is released and reacts with HCO3- to re-form H2CO3, from which CO2 is liberated. The CO2 is then exhaled as a gas.

In addition, some CO2 is directly transported by hemoglobin in the form of carbamate (—NHCOO-). Free a -amino groups of Hb react with CO2 reversibly:

R—NH2 + CO2 ? R—NH—COO- + H+

This reaction is driven to the right in tissues by the high CO2 concentration; the equilibrium shifts the other way in the lungs where [CO2] is low. Thus, carbamylation of the N-termini converts them to anionic functions, which then form salt links with the cationic side chains of Arg a 141 that stabilize the deoxy or T state of hemoglobin.

In addition to CO2, Cl- and BPG also bind better to deoxyhemoglobin than to oxyhemoglobin, causing a shift in equilibrium in favour of O2 release. These various effects are demonstrated by the shift in the oxygen saturation curves for Hb in the presence of one or more of these substances (Figure). Note that the O2-binding curve for Hb + BPG + CO2 fits that of whole blood very well.

Allosteric Effector for Hemoglobin

The binding of 2,3 -bisphosphoglycerate (BPG) to Hb promotes the release of O2. Erythrocytes (red blood cells) normally contain about 4.5 mM BPG, a concentration equivalent to that of tetrameric hemoglobin molecules. Interestingly, this equivalence is maintained in the Hb:BPG binding stoichiometry because the tetrameric Hb molecule has but one binding site for BPG. This site is situated within the central cavity formed by the association of the four subunits. The strongly negative BPG molecule is electrostatically bound via interactions with the positively charged functional groups of each Lys ?82, His ?2, His ?143, and the NH3+-terminal group of each ?-chain.

These positively charged residues are arranged to form an electrostatic pocket complementary to the conformation and charge distribution of BPG. In effect, BPG cross-links the two ?-subunits. The ionic bonds between BPG and the two ?-chains aid in stabilizing the conformation of Hb in its deoxy form, thereby favoring the dissociation of oxygen. In oxyhemoglobin, this central cavity is too small for BPG to fit. Or, to put it another way, the conformational changes in the Hb molecule that accompany O2-binding perturb the BPG-binding site so that BPG can no longer be accommodated. Thus, BPG and O2 are mutually exclusive allosteric effectors for Hb, even though their binding sites are physically distinct.

PHYSIOLOGICAL SIGNIFICANCE OF BPG BINDING

Hemoglobin stripped of BPG is virtually saturated with O2 at a pO2 of only 20 torr, and it cannot release its oxygen within tissues, where the pO2 is typically 40 torr. BPG shifts the oxygen saturation curve of Hb to the right, making the Hb an O2 delivery system eminently suited to the needs of the organism. BPG serves this vital function in humans, most primates, and a number of other mammals. However, the hemoglobins of cattle, sheep, goats, deer, and other animals have an intrinsically lower affinity for O2, and these Hbs are relatively unaffected by BPG. In fish, whose erythrocytes contain mitochondria, the regulatory role of BPG is filled by ATP or GTP. In reptiles and birds, a different organophosphate serves, namely inositol pentaphosphate (IPP) or inositol hexaphosphate (IHP).

Fetal Hemoglobin Has a Higher Affinity for O2

The fetus depends on its mother for an adequate supply of oxygen, but its circulatory system is entirely independent. Gas exchange takes place across the placenta. Ideally then, fetal Hb should be able to absorb O2 better than maternal Hb so that an effective transfer of oxygen can occur. Fetal Hb differs from adult Hb in that the ?-chains are replaced by very similar, but not identical, 146-residue subunits called g -chains (gamma chains). Fetal Hb is thus a 2 g 2.

Recall that BPG functions through its interaction with the ?-chains. BPG binds less effectively with the g -chains of fetal Hb (also called Hb F). (Fetal g -chains have Ser instead of His at position 143, and thus lack two of the positive charges in the central BPG-binding cavity.) Figure compares the relative affinities of adult Hb (also known as Hb A) and Hb F for 02 under similar conditions of pH and [BPG]. Note that Hb F binds 02 at p02 values where most of the oxygen has dissociated from Hb A. Much of the difference can be attributed to the diminished capacity of Hb F to bind BPG; Hb F thus has an intrinsically greater affinity for 02, and oxygen transfer from mother to fetus is ensured.

SICKLE-CELL ANEMIA

In 1904, a Chicago physician treated a 20-year-old black college student complaining of headache, weakness, and dizziness. The blood of this patient revealed serious anemia — only half the normal number of red cells were present. Many of these cells were abnormally shaped; in fact, instead of the characteristic disc shape, these erythrocytes were elongate and crescentlike in form, a feature that eventually gave name to the disease sickle-cell anemia. These sickle cells pass less freely through the capillaries, impairing circulation and causing tissue damage. Further, these cells are more fragile and rupture more easily than normal red cells, leading to anemia.

SICKLE-CELL ANEMIA IS A MOLECULAR DISEASE

A single amino acid substitution in the ?-chains of Hb causes sickle-cell anemia. Replacement of the glutamate residue at position 6 in the ?-chain by a valine residue marks the only chemical difference between Hb A and sickle-cell hemoglobin, Hb S. The amino acid residues at position ?6 lie at the surface of the hemoglobin molecule.

In Hb A, the ionic R groups of the Glu residues fit this environment. In contrast, the aliphatic side chains of the Val residues in Hb S create hydrophobic protrusions where none existed before. To the detriment of individuals who carry this trait, a hydrophobic pocket forms in the EF corner of each ?-chain of Hb when it is in the deoxy state, and this pocket nicely accommodates the Val side chain of a neighbouring Hb S molecule. This interaction leads to the aggregation of Hb S molecules into long, chainlike polymeric structures. The obvious consequence is that deoxyHb S is less soluble than deoxyHb A. The concentration of hemoglobin in red blood cells is high (about 150 mg/mL), so that even in normal circumstances it is on the verge of crystallization. The formation of insoluble deoxyHb S fibres distorts the red cell into the elongated sickle shape characteristic of the disease.3

3In certain regions of Africa, the sickle-cell trait is found in 20 per cent of the people. Why does such a deleterious heritable condition persist in the population? For reasons as yet unkonw, individuals with the trait are less susceptible to the most virulent form of malaria. The geographic distribution of malaria and the sickle-cell trait are positively correlated.

The polymerization of Hb S via the interactions between the hydrophobic Val side chains at position ?6 and the hydrophobic pockets in the EF corners of ?-chains in neighbouring Hb molecules. The protruding "block" on Oxy S represents the Val hydrophobic protrusion. The complementary hydrophobic pocket in the EF corner of the ?-chains is represented by a square-shaped indentation. (This indentation is probably present in Hb A also.) Only the ?2 Val protrusions and the ?1 EF pockets are shown. (The ?1 Val protrusions and the ?2 EF pockets are not involved, although they are present.)

Hemoglobin and Nitric Oxide

Nitric oxide (NO •) is a simple gaseous molecule whose many remarkable physiological functions are still being discovered. For example, NO • is known to act as a neurotransmitter and as a second messenger in signal transduction. Further,endothelial relaxing factor (ERF, also known as endothelium-derived relaxing factor, or EDRF), an elusive hormonelike agent that acts to relax the musculature of the walls (endothelium) of blood vessels and lower blood pressure, has been identified as NO •. It has long been known that NO • is a high-affinity ligand for Hb, binding to its heme-Fe21 atom with an affinity 10,000 times greater than that of O2. An enigma thus arises: Why is NO • not instantaneously bound by Hb within human erythrocytes and prevented from exerting its vasodilation properties?

The reason that Hb doesn't block the action of NO • is due to a unique interaction between Cys93? of Hb and NO • recently described by Li Jia, Celia and Joseph Bonaventura, and Johnathan Stamler at Duke University. Nitric oxide reacts with the sulfhydryl group of Cys93?, forming an S-nitroso derivative:

This S-nitroso group is in equilibrium with other S-nitroso compounds formed by reaction of NO ? with small-molecule thiols such as free cysteine or glutathione (an isoglutamylcysteinylglycine tripeptide):

These small-molecule thiols serve to transfer NO • from erythrocytes to endothelial receptors, where it acts to relax vascular tension. NO • itself is a reactive free-radical compound whose biological half-life is very short (1 - 5 sec). S-nitrosoglutathione has a half-life of several hours.

The reactions between Hb and NO • are complex. NO • forms a ligand with the heme-Fe2+ that is quite stable in the absence of O2. However, in the presence of O2, NO • is oxidized to NO32 and the heme-Fe2+ of Hb is oxidized to Fe3+, forming methemoglobin. Fortunately, the interaction of Hb with NO • is controlled by the allosteric transition between R-state Hb (oxyHb) and T-state Hb (deoxyHb). Cys93? is more exposed and reactive in R-state Hb than in T-state Hb, and binding of NO • to Cys93?precludes reaction of NO • with heme iron. Upon release of O2 from Hb in tissues, Hb shifts conformation from R state to T state, and binding of NO • at Cys93? is no longer favored. Consequently, NO • is released from Cys93? and transferred to small-molecule thiols for delivery to endothelial receptors, causing capillary vasodilation. This mechanism also explains the puzzling observation that free Hb produced by recombinant DNA methodology for use as a whole blood substitute causes a transient rise of 10 to 12 mm Hg in diastolic blood pressure in experimental clinical trials. (Conventional whole blood transfusion has no such effect.) It is now apparent that the "synthetic" Hb, which has no bound NO •, is binding NO • in the blood and preventing its vasoregulatory function.

In the course of hemoglobin evolution, the only invariant amino acid residues in globin chains are HisF8 (the obligatory heme ligand) and a Phe residue acting to wedge the heme into its pocket. However, in mammals and birds, Cys93? is also invariant, no doubt due to its vital role in NO • delivery.

12

Seed Habit, Phylogeny

INTRODUCTION

The seed habit is the most complex and evolutionary successful method of sexual reproduction found in vascular plants. Today, seed plants, gymnosperms and angiosperms are by far the most diverse lineage within the vascular plants. Most of this diversity is accounted for by the angiosperms and Charles Darwin described the rapid rise and early diversification within the angiosperms during the Cretaceous time as 'an abdominable mystery'. Imagine such a plant, producing only one megaspore per megasporangium. The megaspore is not released until the female gametophyte has fully developed within it, complete with archegonium and egg.

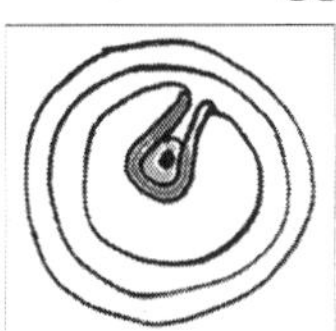

Imagine that the whole megasporangium (with its single megaspore) is released. Imagine too that the megasporophyll and ligule (A, B) extend and enwrap the megasporangium, protecting it. This composite structure is an ovule.

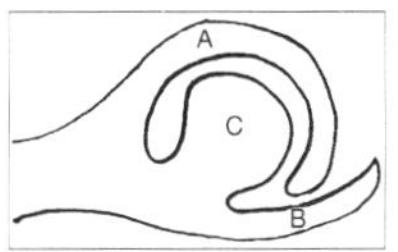

C = megasporangium

Imagine that this protected megasporangium (with its single megaspore and internal gametophyte) is retained on the parent sporophyte until the egg in its archegonium is fertilized and forms an embryo. Only when this has happened, and food reserves are laid down with the embryo, and the outer protective integuments harden, is this structure released. This is what we call a seed.

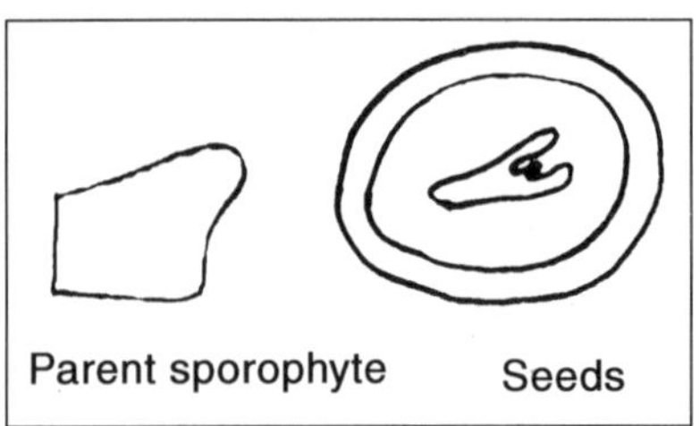

EXISTENCE

The seed plants have an adaptive advantage, occur in a wide variety of habitats and dominate today's flora. This evolutionary success is, at least in part, due to the seed, which is one of the most dramatic innovations during land plant evolution. Morphological evidence for the monophyly of seed plants includes the seed habit itself, but also include vegetative traits like the production of wood by cambium, a secondary meristem. The origin and evolution of the seed habit is a fascinating story by that started in late Devonian times about 385 MYBP. To understand the seed, it helps to think about how it evolved and what it essentially is in terms of origin and function.

The earliest seed plants, "progymnosperms", emerged in the late Devonian. Progymnoperm fossils show vegetative morphologies to seed plants, but not all progymnosperms had seeds or seed-like structures (ovules or pre-ovules). Archaeopteris spp. was the first modern tree, but it sproduced spores rather than seeds. However, it exhibited an advanced system of spore production called heterospory. Heterosporous plants produce two sets of specialized spores: megaspores (haploid female-like megaspores) and microspores (haploid male-like microspores). Heterospory, which probably has been evolved independently in several lineages, is widely believed to be a precursor to seed reproduction.

The progymnosperms are regarded as the ancestors of the seed plants. Fossils of seed-bearing seed ferns (Lyginopteridopsida) exhibit a variety of seed and seed-like structures (see 'The earliest seeds'). 'The seed' might have evolved once or several times during evolution.

Three major evolutionary trends were important for the transition from the seed ferns to the gymnosperms, from the spore to the gymnosperm seed:

1. The evolution from homospory to heterospory and connected with this from megasporangia with many spores to megasporangia with just one functional megaspore.
2. The evolution of the integument, maternal tissue that protects the ovule; the integument forms the seed coat.
3. The evolution of pollen-receiving structures. This includes the transition to water-independence of the pollination/fertilization process (water is required for fern fertilization).

Today there are four major lineages of extant gymnosperm seed plants: Cycadopsida (cycads, "Palmfarne"), Ginkgopsida (ginkgos), Pinopsida/ Coniferopsida (conifers, "Nadelbäume"), and Gnetopsida (gnetophytes). Extinct gymnosperm groups include the Lyginopteridopsida (seed ferns, Pteridosperms, "Samenfarne", paraphyletic group including Devonian/Carboniferous Lyginopterids and Carboniferous/Permian Medullosans), Bennettitales (cycadeoids), Gigantopteridales (gigantopterids), Pentoxylales (*e.g.* Pentoxylon), Caytoniales (*e.g.* Caytonia), Glossopteridales, Voltziales (*e.g.* Emporia), and Cordaitales.

The evolutionary connections between these gymnosperm groups are uncertain and especially the position of the Gnetales is a matter of controverse debate. Based on molecular (DNA sequences), morphological (including fossil seeds), and biogeochemical evidence (oleanone) the different gymnosperm groups are either monophyletic or paraphyletic, and their evolutionary relationships to the angiosperms are unclear. Anthophyte hypothesis: Virtually all morphology-based cladistic analyses propose a monophyletic group consisting of angiosperms + Bennettitales + Pentoxylales + Caytoniales + Glossopteridales + Gnetales (= anthophytes). This implies that the Gnetales are sisters of angiosperms and the gymnosperms are paraphyletic. This hypothesis is not supported by molecular data, which place the Gnetales as sister to all remaining seed plants, and in several molecular analyses even within the conifers as sister to Pinaceae. Based on many molecular analyses, the gymnosperms are a monophyletic sister group of angiosperms.

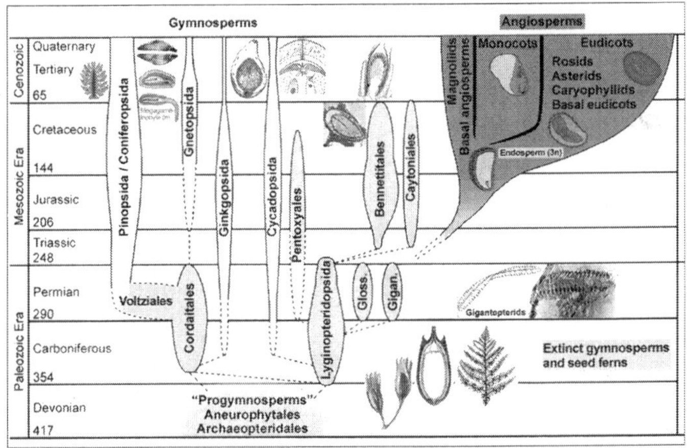

Among several other possibilities, the Caytoniales, Pentoxylales, Glossopteridales, Bennettitales and Gigantopteridales have been proposed as ancestors of the angiosperms. Gigantopterid hypothesis: The Bennettitales (cycadoeoids), which are morphologically highly similar to the Cycadopsida (cycads), are a group of gymnosperms that are highly similar to angiosperms. Oleanane, a triterpene important for insect defence is widespread among angiosperms, has been discovered in fossils of Cretaceous Bennettittales (cycadoeoids) and Permian Gigantopteridales but not in other gymnosperm groups. The gigantopterid hypothesis claims that the angiosperms derived from seed ferns via cycadoeoid-related gymnosperms (gigantopterids).

INSIDE OF SEED

A seed consists of an embryo, stored food and a seed coat. The seed replaces the spore of the seed-less fern plants as propagation, dispersal and deposit/outlast/storage unit.

Ferns and seed plants both exhibit a life cycle in which two heteromorph generations alternate:

1. The dominant diploid sporophyte, which is the fern or spermatophyte plant that you actually see as the 'large-sized' organism in nature.
2. The haploid gametophyte, which for the ferns is the small-sized (few mm to few cm) prothallium of ferns that you might see in nature

In the seed plants the haploid gametophytes became a 'hidden generation' that completely depend on the sporophyte, it was hidden to us until Hofmeister discovered the alternation of generations in seed plants in 1851. In seed plants the male gametophyte (microgametophyte) is hidden in the pollen grain ("Pollenkorn"), and the female gametophyte (megagam-etophyte) is hidden in the ovule ("Samenanlage"), after pollination and fertilization the ovule develops into the seed. Angiosperm and gymnosperm gametophytes and seeds are distinct. To understand the angiosperm seed, it is necessary to understand the gymnosperm seeds, and how it evolved from seed ferns and progymnosperms.

The gymnosperms have 'naked seeds', *i.e.* their ovules and seeds (fertilized ovules) are exposed on the surface of sporophylls and analogous structures. The megagametophyte (female gametophyte, n) develops from the functional megaspore (n) within the nucellus (megasporangium, 2n). The megagametophyte of the gymnosperms is homologous to the megaprothallium (n) of the ferns and is sometimes called primary endosperm (n). The megagametophyte of seed plants is retained and nourished by the parent plant within the ovule. Ovule = megagametophyte + megasporangium (nucellus) + integument (seed coat).

The megagametophytes of the gymnosperms produce several archegonia (n) with egg cells (n). Fertilization by the sperm from the pollen grain (microgametophyte, male gametophye, n) often leads to the development of several embryos (2n) within a single ovule. Polyembryony of gymnosperm seeds is a known phenomenon. In most cases only one embryo survives and therefore relatively few fully developed gymnosperm seeds contain more than one embryo.

In contrast to the gymnosperms ("Nacktsamer"), the angiosperms are "Bedecktsamer", *i.e.* their ovules and seeds are enclosed inside the ovary, which is the base of a modified leaf and is called carpel. Another very important difference to gymnosperms is the angiosperm double fertilization. This leads to an additional novel tissue with maternal protuberance, the triploid endosperm. In mature seeds of most angiosperm species, the embryo is enclosed by endosperm tissue. In addition, angiosperm seeds can be dispersed as fruits, *i.e.* the seeds can have in addition pericarp (fruit coat) around the testa (seed coat). The Triassic and Jurassic age was dominated by gymnosperms, although the first angiosperms evolved during at this time. The rapid rise and early diversification of the angiosperms occurred during the Cretaceous time and was called 'an abdominable mystery' by Charles Darwin.

THE EARLIEST SEEDS

PROGYMNOSPERMS

The earliest seed plants emerged in the Devonian, and have been called "progymnosperms". Progymnoperm fossils show vegetative morphologies to gymnosperms: They were shrubs or trees with laminate leaves, but reproduced by spores like ferns.

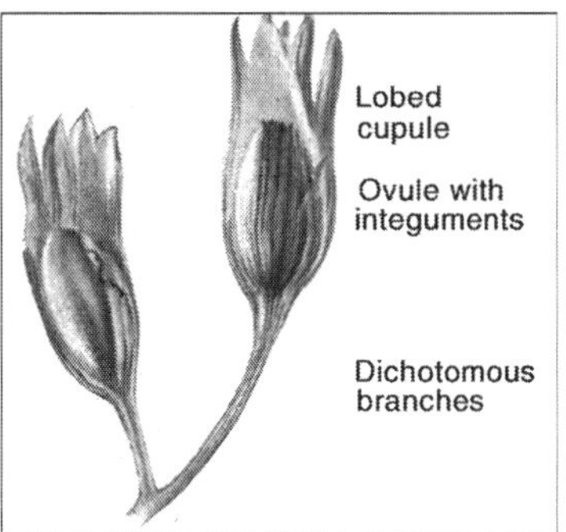

They had a bifacial vascular cambium that produced secondary xylem (wood with pitted tracheids) and secondary phloem. The progymnosperm Archaeopteris spp. was the first modern tree, but it produced spores rather than seeds. However, it exhibited an advanced system of spore production called heterospory.

Heterosporous plants produce two sets of specialized spores: megaspores (haploid female-like megaspores) and microspores (haploid male-like microspores). Heterospory, which has been evolved independently in several lineages, is widely believed to be a precursor to seed reproduction. Thus, both the production of wood and heterospory predate the evolution of the seed. The progymnosperm groups of Archaeopteridales and Aneurophytales are regarded as the ancestors of the seed ferns and seed plants.

DEVONIAN SEED FERNS

The oldest fossils of ovules ("Samenanlagen") or seeds (fertilized, mature ovules) are from the late Devonian (>365 MYBP, Runcaria even 385 MYBP). The earliest seed plants with seeds or seed-like structures are Devonian seed ferns (Lyginopteridposida, Pteridosperms). Several different types of preovules or preovule-like structures are known. Not all characteristics of the seed habit (heterospory, a single megaspore within a megasporangium (nucellus), megasporangium enclosed by an integument, pollen capture before seed dispersal) are evident in these preovules. Runcaria (Gerrienne *et al.* 2004), the oldest seed-bearing seed fern (Middle Devonian, 385 MYBP) had a small, radially symmetrical, integumented megasporangium surrounded by a cupule. The megasporangium bears an unopened distal extension protruding above the multilobed integument. This extension is assumed to be involved in wind pollination (anemophily).

Runcaria sheds new light on the sequence of character acquisition leading to the seed. In general, a seed is simply a mature ovule containing an embryo. An immature ovule consists of a diploid megasporangium (nucellus), containing a single functional megaspore that develops into a haploid megagametophyte. The megasporangium is surrounded by diploid covering layers, the integuments (which evolve into the seed coat).

An integumentary opening at the apical end is important for pollination, wind pollination in seed ferns and modern gymnosperms; this opening evolved into the micropyle.

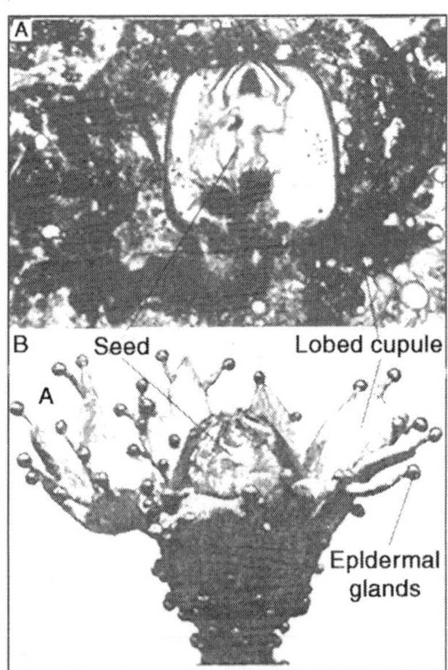

Fig. Fossil (A) and Rconstruction (B) of a Late Devonian Lyginopterid-type Seed of Lagenostoma Lomaxi.

The Late Devonian and Early Carniferous seeds ferns are sometimes collectively called Lyginopterids and include Elkinsia, Archaeosperma,

Lagenostoma, Moresnetia, and Lyginopteris. These earliest seed-bearing seed plants produced their preovules or ovules on dichotomously branched, sterile structures called cupules.

Cupules are cup-like structures that partially enclose the ovule. In these early ovules the nucellus was surrounded by integumentary tissue consisting of more or less free lobes. The integumentary lobes curved inward at their tips, forming a ring around the apical end. The integuments of the ovules evolved through gradual fusion of the integumentary lobes. The integuments later evolved into the seed coat. An opening that was left at the apical end, evolved into the micropyle, it permitted pollen to enter and to fertilize the egg cell. The earliest seed plants employed "hydrasperman reproduction". Pollination occurred when wind-blown pollen was directed into a semi-closed pollen chamber.

In several cases a specialized lagenostome was used for pollen capture. A lagenostome is a funnel-like structure of the nucellus that projects from the top of the megasporangium, it functions as a trumpet-like pollen-trapping device. From there the pollen was delivered to the pollen chamber.

A central column (lagenostome column) was attached to the pollen chamber floor sealed the chamber to provide optimal conditions for pollen germination.

As the megagametophyte matures the opening to the pollen chamber is sealed and the floor is ruptured, allowing fertilization to proceed. The germinated pollen grain (microgametophyte) settles into an intimate proximity to the ovule and delivers the sperm/spermatozoid (has cilia, swims, requires water) into the archegonia with the egg cells.

Fossil records suggest that the sperm delivery required lysis of the megasporangium wall. The important issue was that pollination and/or fertilization became more water-independent during evolution, which facilitated the diversification of seed plants from Carboniferous through to the present day. So far, embryos have not been found in Devonian seed fern fossils.

MEDULLOSAN SEED FERNS OF THE CARNIFEROUS AND PERMIAN

The seed ferns (Lyginopteridposida, Pteridosperms) are a paraphyletic group of extinct gymnosperms. An interesting group, the "medullosan seed ferns" were abundant trees in Carboniferous floodplains end extend well into the Permian. This group includes Trigonocarpus, Pachytesta, Rhynch-osperma, Medullosa and Stephanospermum.

Fossil seeds from medullosan seed ferns (see figures on the left and below) are several mm to several cm long. In some cases even embryo structures have been preserved. The ovules are usually radiospermic, with one end of the integument drawn out into a micropyle that probably helped guide pollen to the megagametophyte within.

Seed-bearing Seed Fern
Emplecopteris Triangularis
from the Permian of China.

A pollination drop mechanism may also have aided pollen capture. A small pollen chamber appears just inside the micropyle. This structure is preserved in detail in a number of Devonian and Carboniferous seed fossils. A loss of the cupule and a loss of the lagenostome column was evident in the medullosan seed ferns.

The integument of Pachytesta and Stephanospermum is three-layered with an epidermis and outer fleshy layer (sarcotesta), covering a though fibrous (sclerotesta) and thin endotesta, which lays adjacent to the nuclleus (see figure on the left). In Stephanospermum konopeonus ovules (Drinnan *et al.* 1990) an apical funnel aided wind-blown pollen capture. Beyond the micropyle and tip of the sclerotestal beak, the sarcotesta flares open as a funnel that tapers to the micropylar opening at the tip of the sclerotestal beak. The rim of the funnel forms he broad apex of the ovule.

The endotesta of Stephanospermum ovules is thin and lines the inner surface of the sclerotesta. The nucellus and the megagametophyte are poorly perserved in the Stephanospermum ovule fossils. In these ovules, the nucellus may have consisted of only a few layers of thin-walled parenchyma surrounding the megaspore membrane. The nuclleus is attached to the integument only at the base of the ovule. The megaspore membrane is robust and consists of a distinctive network of granules and rods of sporopollenin covered by a homogeneous outer layer. Cells of the megagametophyte are not preserved from Stephanospermum ovule fossils.

SEED PHYLOGENY

Phylogenetic relationships among the five groups of extant seed plants are presently quite unclear. For example, morphological studies consistently identify the Gnetales as the extant sister group to angiosperms (the so-called "anthophyte" hypothesis), whereas a number of molecular studies recover gymnosperm monophyly, and few agree with the morphology-based placement of Gnetales. To better resolve these and other unsettled issues, we have generated a new molecular data set of mitochondrial small subunit rRNA sequences, and have analysed these data together with comparable data sets for the nuclear small subunit rRNA gene and the chloroplast rbcL gene.

All nuclear analyses strongly ally Gnetales with a monophyletic conifers, whereas all mitochondrial analyses and those chloroplast analyses that take into account saturation of third-codon position transitions actually place Gnetales within conifers, as the sister group to the Pinaceae. Combined analyses of all three genes strongly support this latter relationship, which to our knowledge has never been suggested before.

The combined analyses also strongly support monophyly of extant gymnosperms, with cycads identified as the basal-most group of gymnosperms, Ginkgo as the next basal, and all conifers except for Pinaceae as sister to the Gnetales + Pinaceae clade. According to these findings, the Gnetales may be viewed as extremely divergent conifers, and the many morphological similarities between angiosperms and Gnetales (*e.g.,* double fertilization and flower-like reproductive structures) arose independently.

PHYLOGENY OF SEED PLANTS BASED ON THREE GENOMIC COMPARTMENTS

DARWIN'S THEORY

Efforts to resolve Darwin's "abominable mystery"—the origin of angiosperms—have led to the conclusion that Gnetales and various fossil groups are sister to angiosperms, forming the "anthophytes." Morphological homologies, however, are difficult to interpret, and molecular data have not provided clear resolution of relationships among major groups of seed plants. We introduce two sequence data sets from slowly evolving mitochondrial genes, cox1 and atpA, which unambiguously reject the anthophyte hypothesis, favouring instead a close relationship between Gnetales and conifers. Parsimony- and likelihood-based analyses of plastid rbcL and nuclear 18S rDNA alone and with cox1 and atpA also strongly support a gnetophyte–conifer grouping. Surprisingly, three of four genes (all but nuclear rDNA) and combined three-genome analyses also suggest or strongly support Gnetales as derived conifers, sister to Pinaceae. Analyses with outgroups screened to avoid long branches consistently identify all gymnosperms as a monophyletic sister group to angiosperms.

Combined three- and four-gene rooted analyses resolve the branching order for the remaining major groups—cycads separate from other gymnosperms first, followed by Ginkgo and then (Gnetales + Pinaceae) sister to a monophyletic group with all other conifer families. The molecular phylogeny strongly conflicts with current interpretations of seed plant morphology, and implies that many similarities between gnetophytes and angiosperms, such as "flower-like" reproductive structures and double fertilization, were independently derived, whereas other characters could emerge as synapomorphies for an expanded conifer group including Gnetales. An initial angiosperm–gymnosperm split implies a long stem lineage preceding the explosive Mesozoic radiation of flowering plants and suggests that angiosperm origins and homologies should be sought among extinct seed plant groups.

The origin of angiosperms has long been considered a fundamental mystery of plant evolution and until recently, the main data available for addressing this question came from morphological and anatomical analysis of living and fossil species, with subsequent cladistic analysis. Morphological homologies, however, are notoriously difficult to ascertain, and many of the relevant characters have been interpreted and reinterpreted many times.

Despite this challenge, a consensus has emerged among morphological cladistic analyses that Gnetales—three bizarre and enigmatic seed plant genera (Welwitschia, Gnetum, and Ephedra)—form a clade with angiosperms and various fossil groups. This "anthophyte" clade is ostensibly the only well-supported relationship among the five main extant seed plant groups and has become the subject of a wide range of evolutionary and molecular developmental studies, several of which have not affirmed a Gnetales–angiosperm relationship.

MOLECULAR PHYLOGENIES

Molecular phylogenies have been largely unable to reach strong conclusions about questions of seed plant evolution other than confirming the monophyly of three of the main groups: Gnetales, cycads, and angiosperms. Published chloroplast and nuclear gene phylogenies, some unrooted, have either potentially supported or, more often, conflicted (23–26) with the anthophyte hypothesis, and there were indications that rbcL at least was saturated at the depth needed to resolve basal seed plant relationships. Several papers using chloroplast rbcL and nuclear 18S reached varied conclusions when different taxa, sequence samples, outgroups, or analysis methods were used, suggesting that the issue was far from settled. We reasoned that optimal resolution of this question may be obtained by using molecular sequences with slower underlying rates of nucleotide substitution than previously utilized, such as sequences from the mitochondrial genome of plants.

Here we have sampled two mitochondrial protein genes, cox1 (cytochrome oxidase I) and atpA (= atp1, ATPase I), from all extant seed plant lineages, including all widely recognized gymnosperm families. We compare phylogenies for these genes to ones based on plastid and nuclear genes for closely matched taxa, developing a comprehensive molecular phylogeny of seed plants on the basis of all three plant genomic compartments. Our results strongly conflict with the anthophyte hypothesis, suggesting instead that Gnetales' closest relatives are conifers, and that the extant sister group to angiosperms is all other seed plants.

METHODS

Twenty-eight new gene sequences were obtained for mitochondrial cox1 and 15 for atpA (29). Approximately 1,416 bp of cox1 and 1,239 bp of atpA were PCR amplified and sequenced. Known or presumed RNA editing sites (with non-synonymous C–T transitions at otherwise conserved amino acid residues) were excluded, as described. Plastid rbcL and nuclear small subunit (18S rDNA) sequences were sampled from the database.

PHYLOGENETIC ANALYSES

Maximum likelihood (ML) analyses of individual data sets were performed by using PAUP, with empirical estimates of base composition. Starting parameters for transition–transversion ratio (ti/tv) and invariant (inv) sites were obtained from initial ML or parsimony (MP) trees and then estimated in formal ML analyses. Gamma, a parameter to account for among-site rate variation, was also estimated in the ML models with inv sites and/or ti/tv held constant. None of the genes exhibited significant heterogeneity in base composition among taxa X^2 test by using PAUP* 4.0. For each gene, at least 10 random input orders were used, and both the Kishino–Hasegawa–Yano (K–H–Y) model for unequal base frequencies and the Felsenstein model were tested; starting parameters and method for ancestral-state reconstruction were also varied; ML topologies were generally insensitive to these choices.

For MP analyses, 250 replicates of random step-wise addition with tree bisection reconnection (TBR) branch-swapping and no weighting were indicated in a heuristic search. Neighbour joining (NJ) was performed with several distance models and parameter values, and the Kimura-2-parameter model with gamma = 0.5 and ML estimates of inv sites were used for analyses shown here unless indicated otherwise. Support for each node was tested with standard bootstrap analysis (BS); 100, 250, and 250 replicates were used for ML, MP, and NJ, respectively. Sequences were also analysed with and without indel regions, and with several different alignments—no significant differences among trees were found. In the final analyses, five small regions of uncertain alignment were excluded from the 18S analyses, and rbcL analyses were restricted to positions 31 to 1,359 from the ATG start codon to include only those portions that were available for Gnetales and other critical taxa.

Detection of Long Branches and Outgroup Selection. Data sets were examined for long branches by using *Relative apparent synapomorphy analysis*.

Initial unrooted analyses were performed with and without putative long-branch taxa. Because rooting of seed plant phylogenies necessarily involves distantly related (non-seed plant) taxa that might easily cause artifactual effects in phylogenetic analyses, analyses were also performed to screen outgroups for suitability in rooted analyses.

Potential outgroups were considered "safe for use" if:

- Their addition did not disrupt ingroup topology when compared with unrooted analyses;
- The assigned outgroup(s) did not cause a decline in tRASA; and
- The assigned outgroup(s) did not result in a significantly long branch according to RASA taxon-variance plot.

RESULT

Although earlier studies offered little hope that a molecular consensus would be forthcoming, it now appears we are rapidly converging on a broadly based well-supported molecular phylogeny of extant seed plants. Our study indicates several important conclusions that differ from those of recent morphological cladistic studies,

- Gnetales are closely related to conifers and, more specifically, they may be derived from within conifers, sister to Pinaceae;
- Rooted phylogenetic trees separate angiosperms from all gymnosperms, implying that extant gymnosperms are monophyletic; and
- Gnetales are unambiguously monophyletic.

In addition, at least four mitochondrial loci several plastid loci, studies of multiple plastid genes and several nuclear genes yield phylogenies inconsistent with an anthophyte clade, with varying taxa and support. Combined evidence from five genes, including mitochondrial *mat*R and *atp*1 (*atp*A), strongly rejects the anthophytes in unrooted trees.

Finally, an unusual derived gene order in the reduced plastid inverted repeat of conifers is also found in Gnetales, a pattern in agreement with a conifer–Gnetales relationship suggested by most of the above studies.

Our analyses showed only minor sensitivity to choice of gene or method of phylogenetic analysis, but these may explain some of the differing results of previous studies by using *rbc*L and nuclear 18S sequences. Apparently both loci have been subject to a greater frequency of multiple substitutions (homoplasy) during seed plant evolution than have the generally slowly evolving mitochondrial *cox*1 and *atp*A genes. This is illustrated by lower consistency index (CI) values in MP analyses.

Given the relatively long branches for both gnetophytes and angiosperms in most of the gene trees, it is probable that these analyses can be affected by long-branch attraction between these groups. Addition of distantly related outgroups that act as long-branch attractors clearly exacerbate the challenge of obtaining a true tree. Here, unrooted analyses, screens for long branches by using RASA, and likelihood analysis with models to account for site-to-site rate variation were used to try to minimize the long-branch attractions.

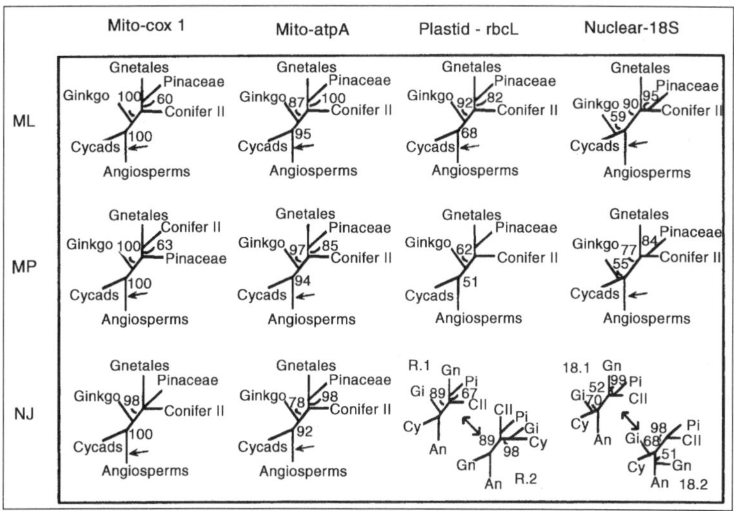

RASA provided an objective means to identify and justify the exclusion of specific long-branch taxa before phylogenetic analysis, and some taxa excluded by RASA did disrupt topologies. For example, the very long-branch *Ephedra* in the *cox*1 data set apparently misled relationships among the three Gnetales genera. If the distant outgroup *Marchantia* is added to *rbc*L analyses, then Gnetales are drawn to the base (not shown).

In other cases, long-branch taxa did not disrupt topology but did alter bootstrap support values. This suggests that studies with distant outgroups must be viewed with caution, and that reanalysis or expansion of earlier seed plant data sets may be fruitful. Finally, multigene multigenome evidence was used to minimize potential gene-specific and compartment-specific effects on phylogeny.

Morphological cladistic analyses have consistently supported a clade with angiosperms and gnetophytes to the exclusion of other extant seed plants, but the molecular hypothesis shown here strongly disagrees. Hypotheses that associate the Gnetales with conifers instead of angiosperms are indicated by all four genes studied, despite marked differences in gene function and genomic location, evolutionary rate, RNA editing, and mode of inheritance.

Unless an unaccounted bias is present in these data or analyses, so that the gene trees do not reflect organismal history, then morphological and other characters that associate gnetophytes with angiosperms in one or more cladistic studies [*e.g.*, double integuments, vessels, lignin biochemistry, tunica formation in the apical meristem, pollen wall structure, and lack of archegonia] probably were derived independently in these two groups. Double fertilization in angiosperms and some Gnetales, of great interest as possibly homologous in the two groups, was probably also separately derived. Alternatively, some of the similarities between Gnetales and angiosperms might be ancestral seed plant traits that have been lost several times during seed plant evolution.

Many earlier (non-cladistic) workers actually considered Gnetales to be close relatives of conifers and convergent with angiosperms; this view was widely accepted for decades. They noted similarities between Gnetales and conifers, including xylem anatomy, cone structure, and the presence of simple linear leaves, inconsistent with the anthophyte hypothesis. Some of these "conifer convergences" may support a Gnetales + Conifer or a Gnetales + Conifer +*Ginkgo* relationship in combined analyses of morphological and molecular data sets.

An unexpected finding is evidence favouring a sister group relationship between Pinaceae and Gnetales, especially in *rbc*L, *atp*A, and four-gene trees. Molecular phylogenies strongly favouring a "gnepines" hypothesis are presented as well by Chaw *et al.* and unrooted five-gene analysis. If correct, these phylogenies would imply that Gnetales are derived conifers that have diverged markedly from their common ancestor with Pinaceae. To our knowledge, this surprising hypothesis has not been suggested by any published morphological cladistic study. Molecular evidence for a Gnetales–Pinaceae clade is decidedly mixed: some studies found a Gnetales–Pinaceae group, but these used *Pinus* as the sole conifer or lacked other major gymnosperm groups. However, the 18S rDNA data clearly favour an alternate hypothesis, with Gnetales sister to all extant conifers.

In fact, the two resolutions of Gnetales represent the only strong disagreement seen in these multigene comparisons. The loss of the large inverted repeat from the plastid genome of *Pinus* and other conifers is not shared by Gnetales; this would require (under the gnepines hypothesis) either independent losses in Pinaceae and Conifer II or the reacquisition of a large inverted repeat in Gnetales. Although the multigene phylogenies give clear evidence that Gnetales and conifers are close relatives and presently support a Gnetales + Pinaceae relationship, these mixed signals suggest that additional data and analyses are needed to evaluate the alternate hypotheses.

13

Indian Agriculture and Biotechnology: The Challenge of the Next Millennium

The world has won the important battle in the area of food security, but the war is still on. A total of 800 million people, that is one out of every six persons, in the developing world do not have access to food. One-third of all pre-school children in the developing countries are food insecure. We, thus, have a big challenge ahead of us as we enter the next millennium. It is true that the mass starvation that was predicted for Asia in seventies and eighties did not occur. It is only because Science was effectively put to work to raise agricultural productivity. In India, the '*Green Revolution*' was a success due to the introduction of improved seeds, fertilizer, irrigation and plant-protection measures combined with positive policy support, liberal public funding for agricultural research and development, and dedicated work of farmers. Notwithstanding all-round achievements, the basic problems of food security, poverty, equity and sustainability, continue to be a cause of concern in India today.

BIOTECHNOLOGY IN INDIA: A PROMISING FUTURE

After becoming an IT bellwether, India is now shifting its focus to the most promising industry of the future, Biotechnology. With its large pool of scientific talent, world-class informationtechnology industry, and vibrant pharmaceutical sector, India is well positioned to emerge as a significant player in the global biotech arena.

Biotechnology is perceived as a revolution throughout the world. Scientists, through Research and Development (R&D), have developed and are continuing to develop cures for diseases that have affected people for decades and even centuries. Scientists recently, have also, clinically developed crops that can withstand the brutalities of weather changes, helping poor farmers of the developing countries to retain their yield and increase their output manifold. Biotech, considered a boon by some, provides great hope to many around the world, and its benefits are and will be realized by more and more people over the years. On the threshold of this new revolution, numerous companies have sprung up to take a piece of the exponentially growing Biotech market worldwide. The ever-decreasing physical boundaries enable biotech companies from the West to tap large markets around the world.

India to this extent holds a good advantage over many other countries of the world. With its large population of over a billion people there is a huge market for products and services. India's population has a very interesting demography that creates almost a perfect environment for biotech companies to shift bases here.

In addition the Indian sub-continent, which occupies only 2.4 per cent of the total global surface area, has the most varied species of flora and fauna. A study shows that, in percentage terms, India has about 7.6 per cent of total mammal species, 12.6 per cent of bird species, 11.7 per cent of fishes and roughly 6.0 per cent of total flowering plants that are present in the world. Biotech companies, by moving to India, can utilize this immense Bio-diversity, can easily find samples and, also conduct field research much more efficiently. Adding to this, India has one of the largest agriculture sectors in the world, and varied climatic zones that can help in research and development of different agribiotech products applicable worldwide

INDIAN ADVANTAGE IN AGRICULTURE

India, today, holds a small share of the global biotech market, but has all the capabilities to become a dominant player. The consumption of biotech products in India is expected to quadruple in the next decade. The human and animal segment of the industry alone is growing by at least 20 per cent.

India has a rich human capital, which is the strongest asset for this knowledge-based industry. India has a large English speaking base and, according to Confederation of Indian Industry estimates, produces roughly 2.5 million graduates in IT, engineering and life sciences, about 650,000 postgraduates and nearly 1500 PhDs qualified in biosciences and engineering each year.

India has proved its competency in selected areas of biotechnology such as, to name a few: capacity in bioprocess engineering, skills in gene manipulation of microbes and animal cells, capacity in downstream processing and isolation methods, and its competence in recombinant DNA technology of plants and animals. India has also allowed assisted stem cell research that permits researchers to use embryos from fertility clinics upon informed consent of the donors, thus giving it a clear head start in this new and promising field in Biotech. Clearly, India has the strength and capabilities in this industry, and a definite advantage to forge ahead and become the chosen location for many biotech companies looking for large markets and low cost qualified manpower to work in their R&D division.

INDIA'S BIO INDUSTRY: SEGMENT REVIEW

The Indian Biotech Industry can be divided into different segments. Following is a review of largest and the fastest growing segments of the Biotech Industry in India.

MEDICAL BIOTECH SEGMENT

The Indian pharmaceutical market is growing very rapidly. According to a study by Mckinsey, Indian Pharma industry is expected to grow to an innovation-led US $25 billion industry by 2010 with a market capitalization of almost US $150 billion from the current US $5 billion generic based drug industry. The vaccine market is expected to grow by roughly 20 per cent.

AGRI BIOTECH SEGMENT

- India being the second largest food producer, offers a hugemarket for biotechnology products, especially agribiotech products.
- India has an excellent scientific infrastructure in agriculture, rich bio-diversity and skilled and low cost human-power.
- In a report by Ernst and Young it is expected that the Nutraceuticals market is roughly US $532-638 million presently and growing.
- With its 8000 kilometer of coastline including Andaman and Nicobar and Lakshwadeep islands, India has a rich aqua culture and its Marine resource development holds great potential.

SERVICES SEGMENT

With increasing number of pharmaceutical companies finding it difficult to conduct entire drug discovery process-in-house they are looking for ways to minimize costs. India has become a very attractive base as the cost of infrastructure is relatively lower compared to other nations.

Foreign companies also benefit from cheaper qualified workforce available in India.

India produces enough qualified graduates each year thus companies looking to expand their operations can easily do so without facing a shortage in labour.

It is estimated that vaccines, contract research, agriculture and human health sectors comprise as much as two thirds of the total market. It is further estimated that health care products would dominate the Indian biotech market, roughly 40 per cent of the total market by the year 2010 followed by agriculture of about 30 per cent.

It is also estimated that contract research and bioinformatics would pick up and account for as much as 25 per cent of the biotech market.

An estimation by CII shows that the Agri-Biotech would see growth rates of as much as 60 per cent, Diagnostic and Therapeutics of about 25 per cent and Vaccines of about 15 per cent. These figures clearly indicate the prospects of the Biotech Industry in India.

GOVERNMENT AND STATE INITIATIVES

The Government of India realized the potential and benefits of this industry at an early stage and formed the Department of Biotechnology in 1986 that has now become the central agency, responsible for policy, promotion of R&D and for international cooperation and manufacturing activities. Some of the initiatives taken by the Government of India are as follows:

The Government of India has been increasing its outlays to provide financial support to this industry. Government of India is also setting up a venture capital fund, to support small and medium enterprises.

Good regulatory framework has been set up for approval of GM crops and rDNA products. Recently, the Government of India decided to make changes to the Drugs and Cosmetics Act to make it more globally compatible.

Indian Patents Bill recently passed by the Parliament allowing 20-year patent term, inline with provisions made by WTO and TRIPS.

India has a sound and widely acknowledged framework of bio-safety guidelines to deal with evaluation, monitoring and release of genetically engineered organisms and there are more than 106 institutional bio-safety committees. The initiatives taken by both Central government and State governments have given a big boost to the Biotech industry in India. Foreign companies looking for new markets and to expand facilities to much more economical locations can find India ever more open and responsive to their needs.

STATE LEVEL INITIATIVES

Many state governments have also realized the benefits/importance of this industry. State governments namely Andhra Pradesh, Tamil Nadu, Karnataka and Maharashtra have begun to formulate their policies, develop R&D centers to encourage and nurture the Biotech Industry. Following is a brief outline of some of the initiatives taken by a few state governments:

Andhra Pradesh

- Government of Andhra Pradesh, in collaboration with the ICICI Limited, has set up a knowledge Park near Hyderabad.
- Development of 'Genome Valley' and Biotech Park with state-of-the art features.
- The Government has initiated many business friendly policies such as 'Single Window Clearance' mechanism geared to relieve the problem of red tape.

Tamil Nadu

- The Government of Tamil Nadu is facilitating in the setting up of the biotechnology enterprise zones (biovalleys) along the lines of Silicon Valley to exploit the bio resources of the state.
- Four state-of-the art biotech parks, a bioinformatics and genome centre will be established, each of which would be leveraging the bioresources of the agro-ecological zones of Tamil Nadu.

Maharashtra

- Has an excellent intellectual infrastructure. Through nearly 1000 institutions, it produces around 163,000 trained technical personnel each year. Some of the best Centers of excellence in India are present in Maharashtra.
- The government is also promoting biotech parks, R&D centers, and pilot plant facilities for underway contract research by putting equity stakes in such projects.

Karnataka

- The Karnataka government has announced a biotech policy to promote this sector and is setting up an institute for bioinformatics in Banglore.
- In addition the state government is also creating a biotechnology fund that will have inflows from the biotech companies. This could be used for incubation of new projects and promotion of the sector in the state.
- Karnataka has planned to launch India's first state sponsored biotechnology venture capital fund to boost their initiatives.

Himachal Pradesh

- Himachal Pradesh has prepared a blue print for promotion of biotechnology industries in the state which includes setting up of biotechnology parks, conservation and exploitation of bio-resources, intensification of R&D, and promoting biotechnology entrepreneurship through tax concessions and relaxed labour laws.
- It is also proposed to provide research based support to the private companies in form of providing for instance, access to a data base of bio-resources which is being developed along with separate entries of endangered medicinal plants.
- The State government has recently announced 100 per cent tax holidays for all biotechnology products up to the year 2012.

INSTITUTIONAL SETUP ANDBUDGETARY ALLOCATIONS IN INDIA

India is one of the first few countries, among the developing countries, to have recognized the importance of biotechnology as a tool to advance growth of agricultural and health sectors as early as in 1980s. India's Sixth Five Year Plan (1980-85) was the first policy document to cover biotechnology development in the country.

The plan document proposed to strengthen and develop capabilities in areas such as immunology, genetics, communicable diseases, etc. In this context, referring to the Council of Scientific and Industrial Research (CSIR), the document suggested to ensure coordination on inter-institutional, inter-agency and on multi-disciplinary basis, full utilization of existing facilities and infrastructures in major areas including biotechnology. Programmes in the area of biotechnology included, tissue culture application for medicinal and economic plants, fermentation technology and enzyme engineering for chemicals, antibiotics and other medical product development; agricultural and forest residues and slaughterhouse wastes utilization and emerging areas like genetic engineering and molecular biology.

The existing national laboratories under the S& T agencies, such as Indian Council of Medical esearch (ICMR) and Council for Scientific and Industrial Research (CSIR) had initiated several esearch programmes to fulfill the above plan objectives. At the top, an apex official agency *viz.* ational Biotechnology Board (NBTB) was set up in 1982, to spearhead development of biotechnology. The NBTB was chaired by Member (Science) of the Indian Planning ommission and had representation of almost all the S&T agencies in the country *viz.* epartment of Science and Technology (DST), Council for Scientific and Industrial Research CSIR), Indian Council of Agricultural Research (ICAR), Indian Council for Medical Research ICMR), Department of Atomic Energy (DAE) and the University Grants Commission (UGC).

NBTB was formed with the specific purpose of the identification of priority areas and for volving a long-term plan for the country in biotechnology as well as to initiate and promote uch activities as conducive for further development of various areas in biotechnology. The BTB issued the " Long Term Plan in Biotechnology for India" in April 1983. This document pelt out priorities for biotechnology in India in view of the national objectives such as self ufficiency in food, clothing and housing, adequate health and hygiene, provision of adequate nergy and transportation, protection of environment, gainful employment, industrial growth and alance in international trade. Later in 1986, NBTB graduated to a full-fledged government epartment called Department of Biotechnology.

At present in India, there are six major agencies responsible for financing and supporting esearch in the realm of biotechnology apart from other sciences. They are Department of cience and Technology (DST), Department of Biotechnology (DBT), Council of Scientific and ndustrial Research (CSIR), Indian Council of Medical Research (ICMR), Indian Council of griculture

Research (ICAR) and University Grants Commission (UGC), Department of cientific and Industrial Research (DSIR). DST, DBT and DSIR are part of Ministry of Science nd Technology while ICMR is with Ministry of Health, ICAR with Ministry of Agriculture and UGC with Ministry of Human Resource and Development. DSIR is the funding agency for CSIR and both of them independently fund biotechnology related research programmes.

Allocations for all of these agencies have gone up in the last decade. Out of this, DBT is the only agency completely devoted to R&D in biotechnology. It is very difficult to estimate the total allocations for this sector per se from other aforementioned agencies as in some cases the allocations are not separately marked as allocations for biotechnology. One faces this kind of constraint especially with those organisations, which are focusing on technological solutions and are not committed for X or Y nature of technology. Thus separately accounting for biotechnology becomes difficult. It would probably become possible only if a detailed survey at the institutional level is undertaken. A broad idea about total allocations by major ndian funding agencies to Science and technology related projects and not necessarily to biotechnology alone. In case of UGC it gives a broad idea not only about S&T related projects but also about other research streams.

DEPARTMENT OF BIOTECHNOLOGY (DBT)

There has been a significant increase in Government of India's outlays for biotechnology over the past decade. Since the time of establishment, in 1986, the allocation for the Department has increased manifold. The budgetary allocations have gone up from ₹ 404 million in 1987-88 to ₹ 1,138 million in 1997-98 and by 2001-2002 it became ₹ 1,863 million. Though the current price allocation figures may not give a complete picture but the budgeted figure for 2002-03 shows a doubling of allocation to ₹ 2,356 million.

In India the developmental allocations are generally made for five years under the National Five Year Plans. The Government has recently finalized the Tenth Five Year Plan (2002-2007). The Working Group on Science and technology for the Tenth Five Year Plan, constituted by the Planning Commission, has proposed an outlay of ₹ 20,750 million for the period 2002 to 2007. This marks a sharp increase of 234 per cent from the budgetary provisions made during the Ninth Plan period (1997-2002) which totaled at ₹ 6,215 million only. The Vision Statement for the Tenth Five Year Plan enumerates the proposal for human genome sequences, proteomics, structural biology and bioinformatics.

The DBT has taken special precaution to find crucial balance between different sections of society as far as technology absorption is concerned apart from promoting industrial development. It is supporting low-cost biotechnology adoption programmes for socially backward communities. The programmes include vermi-composting, use of organic manure, silk-worm rearing, mushroom cultivation, etc. Some training cum demonstration programme are also being supported for them. Efforts are also on for gender mainstreaming. The DBT has launched 11 projects for women in the areas of waste management, bio pesticides, bio fertilizers, floriculture, fish farming for poor women in the rural areas. At the Golden Jubilee Biotechnology Park for Women, industrial modules have been allotted to women entrepreneurs for setting up units in the above-mentioned areas. This Park was inaugurated in 1998 near Chennai the capital of a southern Indian state Tamil Nadu. The Park is located in Siruseri, adjacent to Information Technology Park in an area of about 20 acres. It would have some central facilities for the entrepreneurs for technology resourcing, training and marketing. This is a first unique project of a joint effort by the central and state government in the realm of biotechnology.

CHANGING ROLE OF DBT

In recent times, DBT has taken up a proactive role in promotion of industry. DBT proposed a single window application-processing cell as part of a new regulatory system for the domestic biotechnology sector. The move formed part of the recent recommendations on biotechnology sector made by the Confederation of Indian Industry (CII). Besides recommending setting up of a single window application-processing cell at DBT, CII had also suggested a fixed time frame of 150 days for clearing new biotech proposals.

In this regard, CII had recommended a process whereby a new application would be sent by the single window agency to the Review Committee on Genetic Manipulation (RCGM), which in turn would be required to submit a scientific evaluation report within 60 days of receiving the applications. This report will be submitted to the relevant approval committee, identified by the end product category. For example, in case of agricultural products, it would go to the Genetic Engineering Approval Committee (GEAC), in case of pharmaceutical products to the Drugs and Pharma Approval Committee (DPAC), and in case of food products to the Biotech Foods Approval Committee (BFAC). The GEAC/ DPAC/ BFAC would be required to accord approval or rejection within 90 days of receiving the evaluation report from RCGM. In case of rejection of the application, the applicants will also have the right to appeal to the concerned approval committee. CII had also suggested that any additional information required by RCGM for completeness of the application form would have to be called for within 30 days of receipt of the application. Besides DBT, the recommendations were submitted by CII to 14 other agencies, including seven ministries. Most of the recommendations submitted by CII have been accepted by DBT.

The Council of Scientific and Industrial Research was established in 1942. It is India's largest research and development organization. It has 40 laboratories and 80 field stations/extension centres spread over the length and breadth of the country. The total allocation for CSIR in the year 2000-01 was ₹ 9120 million, which is 13 per cent higher from 1999-2000 (₹ 7940 million). CSIR's Centre for Cellular and Molecular Biology (CCMB) incubated India's first recombinant protein product from a private company, Shantha Biotech, a hepatitis B vaccine, and it has numerous industrial relationships, including a joint venture with Biological E and Amersham Pharmacia to build DNA micro-arrays.

INDIAN COUNCIL OF MEDICAL RESEARCH (ICMR)

Another major institution working in the area of biotechnology is the Indian Council for Medical Research (ICMR) under the Ministry of Health. It is the apex body in the country to promote, coordinate and formulate biomedical and health research. Central Government gives full maintenance grant to the Council, to research in communicable diseases, contraception, maternity and child health, nutrition, non-communicable and basic research.

The total allocation for ICMR from the Central Government (Ministry of Health) was ₹ 1470 million in 2000-01, which was 21 per cent higher over the allocation of the previous year, that is 1999-2000 (₹ 1160 million). The Council is also engaged in research on tribal health, traditional medicine and publication and dissemination of information. In the year 2001 ICMR has launched a major programme in the field of genomics (vector, microbial, human) with the initial allocation of ₹ 510 million. One of the major areas of focus is the disease susceptibility gene identification, especially for, communicable diseases like leprosy, tuberculosis, non-communicable diseases as rheumatic fever or genetic diseases as thalissimia established four centres for developing molecular medicine at All India Institute of Medical Sciences (AIIMS), New Delhi, Sanjay Gandhi Post Graduate Institute of Medical Sciences (SGPGIMS), Lucknow, Jawaharlal Nehru University (JNU), New Delhi and Post Graduate Institute of Medical Education and Research (PGIMER), Chandigarh. Apart from this, ICMR has also established six biomedical informatics units in different parts of the country. It has proposed an allocation of ₹ 1000 million for the Tenth Five Year Plan.

INDIAN COUNCIL FOR AGRICULTURE RESEARCH (ICAR)

In India, agricultural research is being spearheaded by the Indian Council of Agricultural Research (ICAR) under the Ministry of Agriculture. The Council is engaged in conducting research in the field of agriculture, soil and water conservation, animal husbandry, fisheries, dairying, forestry and also agricultural education. The allocation for ICAR from the Ministry of Agriculture was ₹ 13,990 million in 2000-01, which was ₹ 12,060 million in the previous year. It has several research laboratories all over the country conducting research in biotechnology, besides using traditional breeding techniques for different research projects.

ICAR has established a National Research Centre on Plant Biotechnology (NRCPB) at the Indian Agricultural Research Institute (IARI), Pusa, New Delhi, which is fully dedicated to work on plant biotechnology. Annual expenditure over the projects at NRCPB is ₹ 150 million. Apart from this ICAR has also supported two research networks, combining ICAR and even non-ICAR research laboratories to work on crops like rice, cotton, brasica and brinzal. ICAR is also implementing a World Bank supported programme called National Agriculture Technology Programme (NATP) through which huge allocations have been made at different research laboratories for strengthening infrastructure for biotechnological research.

ICAR has also collaborated with DBT and Rockfeller Foundation to jointly launch a National Rice Biotechnology Network (NRBN) in 1988. This project helped in evolving a culture of collaboration among different institutions, which has led to publication of several international papers in established journals. This network put together plant breeders and molecular biologists on the one hand and also provoked interests of private sector in R&D. For instance, Mahyco and Rockfeller Foundation worked together on identifying the relevant genes for saving Indian rice from brown plant hoppers. In this collaborative effort, interaction was not limited to private entities only. It is between private and public sector organisations also. Several universities also have tie ups with private sector organisations. For instance, Tamil Nadu Agricultural University (TNAU) attempted joint trials with Monsanto on weed resistance in soil. TNAU was also partner in monitoring the Bt Cotton field trials of Mahyco-Monsanto alliance. TNAU is also holding in exploratory talks with Rasi seeds (it is also in alliance with Monsanto) to conduct and monitor field trials of Bt crops. Similarly, rice research work at National Research Centre for Plant Biotechnology has attracted business interest of companies like Nath Seeds Ltd. and JK Agri-genetics.

HUMAN RESOURCE DEVELOPMENT AND TRAINING

The National Biotechnology Board had launched an integrated short-term training programme way back in 1984, to cope up with growing demand for highly trained manpower. In the first phase (1984-85), 5 universities were selected for initiating M.Sc./M.Tech programme in this multi-disciplinary area. Subsequently, in 1985-86 and 1986-87, the DBT has added 8 universities/ institutions for M.Sc/M.Tech/Post-doctoral teaching programmes. Subsequently, DBT was entrusted with the responsibility of evolving curriculum for biotechnology courses and meet the demand for human resources in the field of biotechnology. In 1986-87 a model system of post-graduate/post-doctoral teaching in biotechnology in 7 universities/institutions was launched Some of the specialised M.Sc. courses in marine and agricultural biotechnology were launched in 1988-89 at 3 universities. In 1992-93, DBT supported a five year

Integrated Programme in biochemical engineering and biotechnology in Indian Institute of Technology, Delhi and a post-doctoral programme at Indian Agricultural Research Institute, New Delhi.

Now, DBT is supporting 20 M.Sc. courses in general biotechnology, 4 in agricultural biotechnology, one each in medical and marine biotechnology while couple of diploma courses in molecular and biochemical technology 14. The total intake of students in the various postgraduate courses supported by the DBT in the country is around 550 per year. As a part of restructuring of the post doctoral research and training programme, DBT has scraped the on going programme with different institutions and has given this responsibility to Indian Institute of Science (IISc), Bangalore. This is to ensure competitive attitude and quality output in the life sciences. It is being proposed that IISc would award up to 75 fellowships of two-year duration in different streams of biotechnology.

As part of a wider effort for capacity building in institutes of higher learning, full-fledged departments of biotechnology are being set up. The Indian Institute of Science, Bangalore, Indian Institute of Information Technology and Management, Gwalior; and select Regional Engineering Colleges are setting up departments of biotechnology. The All India Council for Technical Education (AICTE) has already approved B.Tech. programmes in biotechnology in eight engineering colleges and has since been advised to develop a model curriculum for undergraduate programmes. All the new departments will have undergraduate, post-graduate and doctoral programmes. Special funding will be provided for the purpose in the Tenth Five Year Plan. Apart from expanding teaching of biotechnology at higher educational institutions a separate module on biotechnology would also be integrated with the school curriculum. The Department of Biotechnology of Government of India will provide the necessary outline of this module so that the National Council of Education Research and Training (NCERT) and the Boards of School Education would be accordingly advised.

Indian University Grants Commission has come out with a scheme to promote higher centres of learning at one place and assist them as much as possible. In this regard, Delhi based Jawaharlal Nehru University (JNU) has been identified by the UGC as centre for excellence in the areas of genomics, genetics and biotechnology. The University has received funds to the tune of ₹ 300 million and is planning to start a new integrated M.Sc./Ph.D programme in life sciences and made to upgrade equipment and library facilities. The new integrated programme in life sciences will reduce the time taken by a scholar to complete his Ph.D by at least within two years. Now the scholar will not be required to undergo a separate M.Phil programme. JNU is aiming at 10 seats for the integrated course and another 20-25 seats in the School of Life Sciences. The University has so far received 40 proposals for possible projects, which can be pursued by them in these fields in the future. Out of the funds that JNU has finally managed to get from the University Grants Commission on its selection as the University with potential for excellence, ₹ 100 million have been set aside for upgradation of facilities and equipments. The rest of the ₹ 200 million would help University to subscribe to some of the 8,000 online journals, both in the filed of Science and social sciences. Besides this, JNU also announced recruitment of more researchers. The tie-ups with industry are also likely to grow. For instance, researchers at the Centre for Biotechnology (CBT) at Jawaharlal Nehru University in New Delhi have been working for four years on a recombinant anthrax vaccine and soon would start Phase I clinical trials in collaboration with Panacea Biotech Ltd. (New Delhi).

PRIVATE SECTOR PARTICIPATION ANDROLE OF FINANCING INSTITUTIONS

In India, biotechnology industry has grown over the past few years at a very rapid pace to reach a nsizeable scale in terms of turnover. According to the available estimates, the size of India's market for biotechnology products could be between US $ 1.5 to US $ 2.5 billion depending upon, among other factors, how a biotechnology product is defined. Of this the agriculture sector market is valued between US $ 450 to US $ 500 million and diagnostic/vaccines market at US $ 150 to US $ 420 million.

AGRICULTURAL BIOTECHNOLOGY

Some of the private foundations such as M. S. Swaminathan Research Foundation (MSSRF), Chennai have taken important initiatives in terms of bridging gap between technology development, its commercialisation and ultimately its diffusion. One of the important project MSSRF launched in early 1990s was establishment of Biovillages in India and China. The Biovillage approach aims at covering principles of ecological sustainability and economic profitability with equity. This project actively promoted interaction between society, industry and R&D institutions. Some of the firms such as Indo-American Hybrid Seeds Company, Bangalore and R&D institution such as Tamil Nadu Agricultural University were the prominent partners. This project boosted the demand for biofertilizers in the Southern Indian Villages.

A SEED INDUSTRY

In the agriculture sector a large number of companies have taken up activities related to biotechnology. Leaving aside subsidiaries of TNCs in India, the agribiotech companies can be classified into three broad groups. One, the larger integrated seed companies which are expanding their R&D to cover biotechnology like Mahyco, Indo-American Hybrid Seed, etc. to develop their own transgenics. Second group is that of smaller companies which have not been active in research or product development but have started employing techniques such as tissue culture for their breeding programmes *e.g.* companies like Kastur Rangan and Adikeshevalu. The third group may cover highly specialized technology companies that undertake services for specified research, like contract research organisations. This is a relatively new concept in the agriculture R&D in India. Some of the companies like Avesthagen qualify in this group.

At this moment all Indian firms are under lot of pressure for technological improvement of their seeds. This pressure emanates largely from their consumers and the growing market penetration by multinational seed corporations. As a result these firms are now exploring the possibilities to embark on biotechnology related research. For instance, a large sized seed company, JK Agri – Genetics has set up a separate division for biotechnology research in those crops in which its owning a larger share in the hybrid seed market. However, on the other side of the spectrum one finds old hands in the field of biotechnology like Indo-American Hybrid Seeds, Bangalore. This was one of the premier companies, which entered in the scene way back in 1992-93 but is still struggling with identification of relevant gene sequences, high capital cost of R&D leading to resource crunch for research and then, on top of that, shortage of skilled manpower.

The company officials point out that despite of so many institutes and universities they are not getting relevant manpower for absorption in R&D units. They are continuing with this urge simply because a breakthrough in biotechnology may help them in retaining their market share in vegetable hybrid seeds as their competitors are gradually tying up with TNCs for accessing their vast pool of gene sequences for crop improvement. Their problem become further confounded from the growing number of relevant genes or gene sequences coming under patent ownership of TNCs.

However, some of the larger companies, which have readily gone for alliances with TNCs, are engaged in back crossing only. For example Maharashtra based Mahyco Seeds which has a tieup with Monsanto has developed transgenic cotton seeds through back-crossing with the genes borrowed from Monsanto for pest resistance. Some companies like Ankur and Rasi, which are not that big in size, have also alliances with Monsanto for similar endeavors. The combined market share of Monsanto through these three tie-ups (in cotton) comes closer to 20 per cent which is next to the public sector National Seeds Corporation (NSC) (45 per cent).

In fact, NSC is also concerned about its declining share over the years and growing concentration in the private sector. In India, almost 8.9 million-hectare of land area are under cotton cultivation with 2.86 million tonnes of cotton lint a year. Since, independence nearly 150 hybrid varieties of cotton have been released and hybrids account for 60 per cent of cotton cultivation. Similarly, small companies such as Bangalore based Kastur Rangan and Adikeshevalu have entered in upscale tissue culture research and related plant development. Actually the tuning point in India's history of seed industry development was the announcement of Seed Policy (1988). This policy statement and related institutional reforms have given a boost to the participation of private sector in the growth of the industry. The private companies have increased their share in the industry from 20 per cent in 1981 to 73 per cent in 1997 and 76 per cent in 2001. There are nearly 150 registered seed companies. The market size for seeds has grown from ₹ 10,000 million in 1994-95 to ₹ 22,000 million in 1998-99. The interesting point to be noted is that the concentration in favour of organized private sector has grown many folds. The share of public sector seed supply has declined from 40 per cent to 25 per cent while share of unorganized sector has declined from 25 to 15 per cent in the same time period. The share of organized seed sector has grown from 35 per cent to 60 per cent.

BIOFERTILIZERS AND BIOPESTICIDES

In recent past, private sector participation in production of biofertilizers has grown at a very high pace. The production has gone up from 2.5 tonnes in 1992-93 to 10.38 tonnes in 1999-00. Accordingly, the number of firms engaged in biofertilizers has also gone up from 35 in 1992-93 to 95 in 1999-00. The Biofertilizers Statistics being regularly brought by the Fertilizer Association of India shows that the production of fertilizers is concentrated in Western Indian states.

They are Gujarat (3 units), Madhya Pradesh (7 units), Maharashtra (24 units) and Rajasthan (3 units). However, some of the Southern states like Tamil Nadu have also encouraged private industry to set up biofertilizer units. In three years time more than 13 units have come up in the state. There are several types of biofertilizers being marketed in India. Some of the prominent ones are Rhizobium, Azatobacter and Azospirillum. The Indian Council for Agriculture Research and Department of Biotechnology have actively encouraged application of rDNA technology for better quality Rhizobium and Azatobacter.

In order to help industry, DBT has established certain repositories to keep micro organisms. In case of biofertilizers, the established repository for microbes is "National Facility for Rhizobium Culture Collections", Division of Microbiology, IARI, New Delhi. The others are National Centre for Conservation and Utilisation of Blue-Green Algae, IARI, New Delhi, Microbial Type Culture Collections, Institute of Microbial Technology, Chandigarh, National Facility for Marine Cyanobacteria, Bharatidasan University, Tiruchurapalli, Facility for Mycorrhizal Culture Collections, Tata Energy Research Institute, New Delhi. However, its important to mention here that the demand for biofertilizer suffers from three major factors: poor and uneven quality, short shelf life and small contribution of crop yield for biopesticides. DBT established a Biocontrol Network Programme in 1989. The major emphasis in this programme is to develop better formulations and cost effective commercially viable biopesticides including microbial pesticides, parasitoids and bacteria for use under IPM. The project has 80 R&D projects and total allocations of 1.8 million.

MEDICAL BIOTECHNOLOGY

The companies in medical biotechnology in India can be divided into three broad categories. One is that of small startup companies that have indigenously developed biotech products, *e.g.*, Shantha Biotech and Bharat Biotech.

Then there are large companies, which have started responding to biotechnology and have in fact incorporated biotechnology in their work plan, for instance, Dr. Reddy's Laboratory (DRL), Ranbaxy Laboratories and Wockhardt Ltd. The third group has start-ups, which are all set to emerge as contract research organisations (CROs). Largely their work comes from TNCs.

Then there are companies like Biocon India which may not fit well in this kind of classification as they have an established presence in the industrial biotechnology (the fermentation sector) and a growing presence in the pharmaceutical sector, so eventually encompany our first and second category. Biocon set its sights on biopharmaceuticals and using its capabilities in a wide range of fermentation technologies by 1995.

Two years on, Biocon established Helix – a wholly owned subsidiary to develop its biopharma operations, which began with a range of anti-cholestrol statin drugs. Another subsidiary of this company, Clinigene International, was set up to initiate longitudinal clinical studies in select disease segments. Thus there is a synergy-based expansion, Biocon is now recognised as the country's leading biotech conglomerate.

SHANTHA BIOTECH

This is one of the leading examples of our first category of firms – small start ups with their own biotech products. Shantha Biotech has the credit of developing India's first world class HepatitisB vaccine and making it available at one third of the prevailing market price of imported vaccines. This company has an active biotechnology programme since 1994. Now Pfizer Ltd., the major Pharma TNC, has obtained the first refusal rights from the Hyderabad based Shantha Biotechnics Pvt. Ltd. for exclusively marketing the products to be developed by the latter in future. Earlier, Pfizer had entered into a co-marketing agreement with Shantha Biotech for marketing the latter's recombinant DNA vaccine for Hepatitis-B. Shantha Biotech is currently in advanced stages of discussion with one European Pharma major and about three US based pharma companies for its research projects.

The company plans to research on protein purification, molecular cloning and expression of native and synthetic genes. Shantha Biotech will also be offering polyclonal and monoclonal antibody development and formulation of certain types of vaccines. Shantha Biotech has developed in-house expertise in recombinant DNA technology and are very strong in development of cell lines for development of recombinant products. The company has invested ₹ 100 million in the biotech facility with external funding (from Bank of Oman).

DR REDDY'S LABORATORY (DRL)

Some of the companies like DRL have grown in the recent past. It was established in 1984. DRL is a big pharmaceuticals company and now is setting up biotech production facilities as per the US FDA specifications. The company has also identified biogenerics as significant market area which is estimated at $14 billion. It has pegged its brand value at ₹ 9,160 million as against last year's value at ₹ 2,850 million.

The biotechnology facility includes setting up of three class 10,000 bulk recombinant protein production suites and new formulations facility as well. The biotechnology business would cover therapeutics, vaccines and diagnostics. DRL has a research alliance with Centre for Cellular and Molecular Biology (CCMB) Hyderabad. DRL has also established a research subsidiary in Atlanta called Reddy US Therapeutics Inc., as well as a contract research subsidiary that will focus on genomics.

RANBAXY LABORATORIES

Ranbaxy, India's largest Pharma company with sales of more than $500 million in the year 2000, also views innovation in biotechnology as key to its future. It is one of the oldest post-independence firm founded in 1968. The company has branched out from creating new formulations of existing drugs and has half a dozen molecules under development. Ranbaxy has collaborations with several U.S. and European companies to develop new formulations and technologies. For example, Ranbaxy and Vectura Ltd. (Bath, U.K.) announced in 2001 that the Indian company's Ranbaxy B.V. subsidiary (the Netherlands, Antilles,) will develop oral formulations using Vectura's controlled release drug delivery technology, with Ranbaxy providing clinical development, scaleup, manufacturing and marketing expertise. Ranbaxy has set a model in India in terms of drug development. The model suggests that develop a molecule upto the level a domestic company can afford to do generally upto the first phase of the clinical trial and then outsource it to a leading MNC for further development and later on explore the possibilities for marketing tie-ups.

WOCKHARDT LTD.

Wockhardt Ltd. established nearly four decade ago, is the fifth largest pharmaceutical company of India. Wockhardt has grown at over 20 per cent annually for the last 5 years. In the year 2000, it acquired Merind from the

Tata's and besides Wallis Laboratories, a company in the UK, apart from entering into marketing alliance with the American Sidmak Labs. The company is formulating its biotechnology strategy around these initiatives. It has decided to split its business into two separate entities. Wockhardt Ltd. will remain as a pharmaceutical company while the new entity would focus on Life sciences only.

The new company (the name of which has yet to be finalized) will have all the other businesses namely hospitals, nutrients, IV fluid and agrovet (Crop protection). The R&D activity is also being categorized into three divisions. The first division concentrates on developing new bulk drugs (Novel drug delivery system (NDDS)) and generics. The second division concentrates on recombinant biotechnology. At present, the second division is focusing on technology absorption with the assistance of scientists from abroad. The third division is dedicated for discovery of new molecules. Wockhardt is expected to come out with its first investigational new drug (IND) in anti-infective therapeutic segment by end of 2002.

AGRICULTURE IN INDIA

India's record of progress in agriculture over the past four decades has been quite impressive. The agriculture sector has been successful in keeping pace with rising demand for food. The contribution of increased land area under agricultural production has declined over time and increases in production in the past two decades have been almost entirely due to increased productivity.

Contribution of agricultural growth to overall progress has been widespread. Increased productivity has helped to feed the poor, enhanced farm income and provided opportunities for both direct and indirect employment. The success of India's agriculture is attributed to a series of steps that led to availability of farm technologies which brought about dramatic increases in productivity in 70s and 80s often described as the Green Revolution era. The major sources of agricultural growth during this period were the spread of modern crop varieties, intensification of input use and investments leading to expansion in the irrigated area. In areas where 'Green Revolution' technologies had major impact, growth has now slowed.

New technologies are needed to push out yield frontiers, utilize inputs more efficiently and diversify to more sustainable and higher value cropping patterns. At the same time there is urgency to better exploit potential of rainfed and other less endowed areas if we are to meet targets of agricultural growth and poverty alleviation. Given the wide range of agroecological setting and producers, Indian agriculture is faced with a great diversity of needs, opportunities and prospects. Future growth needs to be more rapid, more widely distributed and better targeted. These challenges have profound implications for the way farmers' problems are conceived, researched and transferred to the farmers. On the one hand agricultural research will increasingly be required to address location specific problems facing the communities on the other the systems will have to position themselves in an increasingly competitive environment to generate and adopt cutting edge technologies to bear upon the solutions facing a vast majority of resource poor farmers.

In the past agriculture has played and will continue to play a dominant role in the growth of Indian economy in the foreseeable future. It represents the largest sector producing around 28 per cent of the GDP, is the largest employer providing more than 60 per cent of the jobs and is the prime arbiter of living standards for seventy per cent of India's population living in the rural areas.

These factors together with a strong determination to achieve self-sufficiency in food grains production have ensured a high priority for agriculture sector in the successive development plans of the country.

An important facet of progress in agriculture is its success in eradication of its critical dependence on imported foodgrains. In the 1950's nearly 5 per cent of the total foodgrains available in the country were imported. This dependence worsened during the 1960's when two severe drought years led to a sharp increase in import of foodgrains. During 1966 India had to import more than 10 million tonnes of foodgrains as against a domestic production of 72 million tonnes. In the following year again, nearly twelve million tonnes had to be imported. On the average well over seven per cent of the total availability of foodgrains during the 1960s had to be imported. Indian agriculture has progressed a long way from an era of frequent droughts and vulnerability to food shortages to becoming a significant exporter of agricultural commodities. This has been possible due to persistent efforts at harnessing the potential of land and water resources for agricultural purposes. Indian agriculture, which grew at the rate of about 1 per cent per annum during the fifty years before independence, has grown at the rate of about 3 per cent per annum in the post independence era.

AGRICULTURE – SUB-SECTORS

Indian agriculture broadly consists of four sub-sectors. Agriculture proper including all food-crops oilseeds, fibre, plantation crops, fruits and vegetables is the largest accounting for nearly 70 per cent of the agriculture sector as a whole. The rapid growth in this sub-sector through exploitation of wastelands and fallows, spread of irrigation and adoption of production enhancing technologies was critical in transforming India from a country vulnerable to food shortages to one of exportable surplus. Although this sub-sector has made impressive progress its share in the sector as a whole has declined from 78 per cent in 1960-61 to less than 70 per cent by early 90s.

Table. Agriculture sub-sectors – value of output (%)

Sub-sector	1960-61	1970-71	1980-81	1993-94
Agriculture	78.0	79.3	75.3	68.5

Livestock	16.6	15.1	17.2	24.5
Forestry	4.4	4.4	5.9	3.8
Fisheries	1.0	1.2	1.6	3.2
	100.0	100.0	100.0	100.0
Total value at Current price (₹ Billion)	89	218	615	2740

Correspondingly the share of livestock sector has increased considerably. The livestock industry has grown from ₹ 15 billion in early 1960s to ₹ 100 billion by 1980-81 and ₹ 672 billion by 1993-94. In nominal terms the sector grew at almost 15 per cent per annum during 1980s. Milk production, which was almost stagnant for two decades ending 1970, grew by over 5 per cent per annum in the 80s. Similarly, production of eggs increased at the rate of about 6.5 per cent during the same period. As a result the share of livestock increased from about 17 per cent till early 80s to 25 per cent by 1993-94.

Though it plays relatively a minor role within the sector as a whole, fishing sub-sector activities have been on the rise. The sub-sector has grown from only ₹ 3 billion in 1970-71 to nearly ₹ 90 billion in 1993-94. The growth was particularly rapid in 70s and 80s. Value added increased at over 5 per cent per annum during this period. In real terms forestry and logging activities have been on the decline since mid seventies. As of 1993-94, the size of the industry in terms of value of output was 103 billion.

Table. Foodgrains Production

Year	*Area (million ha)*	*Production (million tonnes)*	*Yield kg/ha*
1950-51	97.32	50.82	522
1960-61	115.58	82.02	709
1970-71	124.32	108.42	872
1980-81	126.67	129.59	1023
1990-91	127.84	176.39	1379
1998-99	124.0	195.25	1574

Over the past three decades, the country has successfully transformed itself from a food deficit economy to one which is essentially self sufficient in availability of foodgrains and other essential commodities, albeit only at the prevailing level of effective demand. Annual aggregate foodgrains production, which averaged about 82 million tonnes in 1960-61 increased to 123.7 and 172.5 million tonnes for the trienniums ending 1980-81 and 1990-91 respectively. Current (1998-99) production level is 195 million tonnes and the country has been able to accumulate substantial, (35 million tonnes) stocks of foodgrains to cope up with any sudden difficulties arising from drought or a similar situation in any part of the country.

Increased outputs, have been achieved chiefly by adopting, since mid sixties, a strategy aimed at increasing foodgrains production by concentrating public sector efforts and resources in regions with a high potential for quick and substantial productivity gains through increased cropping intensity and average yields. These were the areas favoured by agroclimatic resource conditions and where irrigation facilities already existed or could be developed relatively rapidly. The main elements of this strategy were:

(i) expansion of irrigation coverage,

(ii) increased provision and utilization of key inputs – mainly high yielding varieties (HYVs) of crops, mainly of wheat and rice and chemical fertilizers and plant protection chemicals,

(iii) expansion and improvement of institutional support services such as research and extension and

(iv) price policies favourable to producers of major foodgrains.

The success of this strategy was made possible by development and availability of replicable production technology packages, so called 'Green Revolution' technologies. Irrigation facilitated double cropping and widespread adoption of HYVs. The HYVs performed particularly well under irrigated conditions, were highly responsive to fertilizers and their short duration permitted increases in cropping intensities.

Irrigation development was the cornerstone of the strategy. Undivided India was amongst the largest irrigated areas in the world. With partition nearly one-third of the irrigated area went to Pakistan. At the time of independence the net irrigated area was 20.9 million ha (gross irrigated area 22.6 million ha). Recognizing large-scale development of irrigation facilities as critical to rapid agricultural growth, the country has spent about ₹ 45,000 crores on irrigation development in the first four decades after independence. During the period 1950-51 to 1965-66 development of irrigation through government canals grew from 7.2 million ha to 9.8 million ha – a growth rate of 2.1 per cent per annum. During 1970s this pace dropped slightly to 1.9 per cent. In 1980s the rate of increase dropped significantly to 1.1 per cent per annum.

The growth of tube-well irrigation, however, increased rapidly from 4.5 million ha in 1970-71 to 9.5 million ha in 1980-81 and then to 14.3 million ha by 1990-91. The net irrigated area increased from 31 million ha in 1970-71 to 53.5 million ha in 1995-96 which corresponds to 22 per cent of the net sown area in 1970-71 and 37.63 per cent in 1995-96 *(Tables 3, 4).* With improvements in irrigation efficiency the gross irrigated areas has increased to 71.51 million ha. The percentage of gross cropped area service by irrigation increased from 18.3 per cent in 1960-61 to 23.0 per cent in 1970-71 and to over 38 per cent at present.

Table. India – irrigated area

Year	*Net*	*Gross*	*Net as % of net sown*	*Gross as % of gross cropped*
	——— *million ha* ———		———— % ————	
1970-71	31.10	38.19	22.17	23.04

1980-81	38.72	49.78	27.66	28.83
1990-91	48.02	63.20	33.61	34.03
1995-96	53.51	71.51	37.63	38.33

Table. Sources of irrigation

Year	*Net Irrigated Area (million ha)*	*Sources*				
		Canals	*Tanks*	*tube-wells*	*Other wells*	*Other Sources*
		(% net irrigated area)				
1960-61	24.80	10.4	4.6	0.2	7.2	2.4
1970-71	31.10	41.28	13.22	14.34	23.88	7.29
1980-81	38.72	39.49	8.22	24.62	21.08	6.59
1990-91	48.02	36.34	6.13	29.69	21.73	6.11
1995-96	53.01	32.04	5.81	33.52	22.16	6.47

Fertilizers have constituted yet another key input in addition to expanded irrigation and spread of HYVs in achieving goals of high production and productivity. India currently occupies third position in the world, after China and USA, in terms of fertilizer production and consumption. Consumption of fertilizers has increased from 1.54 million tonnes in 1967-68, representing the pre green revolution era, to 17.31 million tonnes currently (1997-98).

The average per hectare use of fertilizers currently around 85 kg per hectares is the lowest among several Asian countries.

However, rice and wheat account for a major fraction, around 65 per cent of the total fertilizer consumed in the country, with very little fertilizers going to the rainfed areas.

According to some current projections, fertilizer's use will need to increase to 30-35 million tonnes to meet the foodgrains need of 2020. The demand for nutrients will stretch by almost another 15 million tonnes if requirements for horticulture, vegetables and plantation and commercial crops are included. At present domestic production of N and P fertilizers (13.42 million tonnes) falls short of consumption by over 20 per cent. In addition the entire requirement of K fertilizer is imported.

Most agricultural development programmes initiated in 1960s were concentrated in regions of high potential. Thus five states, Punjab, Haryana, Uttar Pradesh, Andhra Pradesh and Tamil Nadu account for 50 per cent of the country's net irrigated and 53 per cent of the gross irrigated area. The combination of expanding irrigation coverage and widespread adoption of short duration HYVs led to significant increases in cropping intensities. Acreage cropped more than once per year increased from 13 million ha in 1950-51 to about 44 million ha at present.

Average cropping intensity for the country as a whole rose from 115 per cent in 1960-61 to 131 in 1993-94. By 1993-94 cropping intensity has risen to 187 per cent in Punjab, 167 per cent in Haryana and 142 per cent in Uttar Pradesh.

Table. Cropping intensity

Years	Net sown area	Gross cropped area	Area cropped more than once	Cropping intensity (%)
	million ha			
1950-51	118.8	131.9	13.1	111.1
1960-61	133.2	152.8	19.6	114.3
1970-71	140.3	165.8	25.5	118.2
1980-81	140.0	172.6	32.6	123.3
1990-91	143.0	185.7	42.7	129.9
1993-94	142.0	186.4	44.4	131.2

An important consequence of the strategies adopted since sixties has been to boost production of, chiefly, two crops rice and wheat. Their share in total foodgrains production went up from 57 per cent in 1970-71 to more than 75 per cent in 1990-91. Production of foodgrains other than rice and wheat did not increase significantly and in the eastern region even the yield of rice did not increase.

Agricultural production and income rose substantially in the north-western states of Punjab, Haryana, Western Uttar Pradesh, parts of Rajesthan, Tamil Nadu and Andhra Pradesh. By contrast productivity and output growth have been modest in eastern and central India and in deccan plateau. Progress was particularly slow in rainfed areas, which account for over 60 per cent of the cropped area and where a great majority of rural poor are concentrated. An important impact of the strategies pursued in the 'Green Revolution' period has been intensification of regional disparities and imbalances in agricultural development and food availability and hence levels of food security. For the country as a whole while per capita availability of cereals has increased substantially, that of pulses has decreased significantly.

In summary an annual increase in foodgrains production of 3.22 per cent during fifties was mainly because of expansion in area. Sixties recorded a low annual growth rate of 1.72 per cent necessitating large-scale imports of foodgrains.

Table. Rice and Wheat Production

Production (million tonnes)

Year	Rice	Wheat	Foodgrains	R+W as % of Foodgrains
	million tonnes			
1970-71	40.80	20.86	108.4	57.4
1980-81	49.91	34.55	129.6	65.4
1990-91	72.78	53.03	176.4	75.0
1996-97	81.74	69.35	199.4	75.0

Annual growth of 2.08 per cent was recorded during seventies. This decade was the turning point in India's foodgrains economy leading to self-sufficiency through significant productivity increase first in wheat and later in rice in the eighties. An annual growth of 3.5 per cent in foodgrains in eighties was the hallmark of green revolution that enabled India to become self-sufficient and even a marginal exporter. The pace of growth slowed in nineties barely making or even slower than the population growth rate. This is a matter of concern.

Table. Annual compound growth rate of foodgrains production

Crop	1950-51 to 1959-60	1960-61 to 1969-70	1970-71 to 1979-80	1980-81 to 1989-90	1990-91 to 1997-98
Rice	3.28	- 8.05	1.91	4.29	1.53
Wheat	4.51	5.90	4.69	4.24	3.67
Coarse cereals	2.75	1.48	0.74	0.74	- 0.49
Total cereals	3.00	2.51	2.37	3.63	1.84
Pulses	2.72	1.35	- 0.54	2.78	0.76
Total foodgrains	3.22	1.72	2.08	3.54	1.66

Table. Per capita net availability of cereals and pulses per day (g)

Year	*Cereals*	*Pulses*	*Total*
1951	334	61	395
1961	400	69	469
1971	418	51	469
1981	417	38	455
1991	468	42	510
1998	451	33	484

HORTICULTURE

The diversity of physiographic, climate and soil characteristics enables India to grow a large variety of horticultural crops – fruits, vegetables, flowers, spices, aromatic and medicinal plants, plantation crops etc.

India is the largest producer of fruits in the world and second largest producer of vegetables. The area under fruits estimated at 1.45 million ha in 1970-71 grew to 2.8 million ha in 1991-92 and then more rapidly to 5 to 6 million ha by 1994-95.

This sector is likely to grow rapidly in the future both on account of internal demands and export opportunities.

ANIMAL HUSBANDRY AND FISHERIES

Animal husbandry and dairying sub-sector plays an important role in overall economy and in social development. The contribution of the sub-sector is estimated to be about 25 per cent of the total value of output of agricultural sector.

Table. Production of milk, eggs and fish in India

Year	Milk production/ availability million tonnes/annum	g./day	Fish (000 tonnes)	Eggs (million)
1950-51	17.0	124	752	1832
1960-61	20.0	124	1160	2881
1970-71	22.0	112	1756	6172
1980-81	31.6	128	2442	10060
1990-91	53.9	176	3836	21101
1998-99	75.0	210	5388	28400

The sector also plays a significant role in supplementing family incomes and generating employment in the rural sector particularly among the landless, small and marginal farmers and women besides providing nutritious food. Production of suitable cross breeds and their wider adoptions has contributed to increasing country's milk production, which has now reached 75 million tonnes annually. Similarly, genetic improvement and better management practices have markedly pushed up production of poultry and eggs.

Table. Number of holdings (million)

Category of Farmers	1970-71	1980-81	1990-91
Marginal	35.7 (50.6)*	50.1 (56.4)	62.1 (59.0)
Small	13.4 (19.0)	16.1 (18.1)	20.0 (19.0)
Semi-medium	10.7 (15.2)	12.5 (14.0)	13.9 (13.2)
Medium	7.9 (11.2)	8.1 (9.1)	7.6 (7.3)
Large	2.8 (3.9)	2.2 (2.4)	1.7 (1.6)
All groups	70.5 (100)	89.0 (100)	105.3 (100)

Through the overall contribution of fisheries sub-sector is small, development of prolific and fast growing forms of several common fish species coupled with breakthrough in breeding under captivity, fish seed production and multi-layer fish culture has resulted in registering a very high annual growth of 11 per cent in aquaculture production during the past decade.

Table. Area operated - % of total area

Category of Farmers	1970-71	1980-81	1990-91
Marginal	9.0	12.1	14.9
Small	11.9	14.2	17.3
Semi-medium	18.5	21.1	23.1
Medium	29.7	29.6	27.2
Large	30.9	23.0	17.4
All groups (million ha)	162.1 (100)	163.8 (100)	165.6 (100)

SUSTAINABILITY CONCERNS

Several indicators highlight increasing concerns of sustainability in areas which have largely contributed to increased production in the 'Green Revolution' era. Adoption of high yielding cultivators is virtually complete. Almost entire wheat and rice crops in the states of Punjab, Haryana and Western Uttar Pradesh are irrigated. In the higher production regions yields are plateauing and most traditional sources of productivity growth having been exhausted future gains in production have to come from elsewhere.

At farmers' level concerns are being expressed in several ways. Many farmers believe that the input levels have to be continuously increased in order to maintain high yields. In sixties and seventies most farmers used only nitrogenous and phosphate fertilizers to achieve high yields. Due to widespread deficiencies of several secondary and micronutrients, most farmers now have to apply higher doses and a greater variety of fertilizers to maintain crop yields.

Results from many long term studies on rice-wheat cropping system show a declining yield trend when input levels were kept constant – thus the growth rate of system productivity has been declining relative to growth rate of nutrients use.

Lowering of groundwater tables due to intensive rice-wheat system in many areas is resulting in increased costs of lifting water in the intensively cultivated high production areas, diseases and pest problems are turning more serious than ever before and pose both short and long large problems. It is reported that some weeds have developed resistance to the commonly used herbicides. What this implies is that the farmers are applying increasing amount of herbicide incurring increasing cost without the benefit of effective control. Pesticide residues entering the food chain and overall safety in use of pesticides continue to be serious problems.

Other emerging problems threatening sustainability of intensive cropping system *e.g.* rice-wheat include loss in biodiversity related issues. Large areas planted to a single/few varieties of a crop is a potential cause of concern. As the diversity is reduced natural processes that control and affect habitat quality and genetic expression weaken and for this reason internal and natural control mechanisms must be replaced by more externally applied artificial controls in the form of management and inputs which in due course lead the system towards unsustainability.

Groundwater is the major source of meeting the irrigation needs of irrigated agriculture. Currently about half the area under irrigation in the country is irrigated from groundwater sources.

Large-scale groundwater development has led to fall in the water table in many areas. Over pumping is leading to declining water table levels and failure of tube-wells. Pumping costs are increasing, as is the energy consumption. In the coastal areas this has led to ingress of sea water, with serious environmental implications.

Changes in water quality are adversely affecting agriculture and vice-versa. Inefficient and/or over use of fertilizers and pesticides in agriculture and untreated disposal of industrial and urban wastes are leading to increasing contamination by such elements as lead, zinc, copper, chromium, cadmium particularly in areas having high industrial activity *e.g.* in districts of Ludhiana, Faridabad, Kanpur, Varanasi etc.

An increase in the content of arsenic has been reported in several of the districts of West Bengal. This is attributed amongst other causes to the lowering of groundwater table due to excessive groundwater withdrawal and is leading to serious and widespread toxicity problems adversely affect the health of hundreds of thousands people of the region.

CHANGING LAND-USE AND FUTURE OF AGRICULTURE

One of the most important consequences of growing pressure on land is the declining trend in the average farm size and the pattern of holdings. According to the latest Agricultural Census in 1970-71 there were 70 million holdings operating 162 million ha.

By 1990-91 there were 105 million holdings operating 165 million ha. The average farm size decreased from 2.30 ha in 1970-71 to 1.57 ha in 1990-91. As of 1990-91 about 78 per cent of holdings were small (1.0 to 2.0 ha) and marginal (<1.0 ha). A little more than 20 per cent of the farmers were semi-medium (2.0 to 4.0 ha) and medium (4.0 to 10.0 ha).

Large farmers (>10.0 ha) constituted only 1.6 per cent of the total holdings. Over the twenty-year period since 1970 the proportion of marginal farmer has increased from 50 to 59 per cent and that of large farmer has declined from about 4 to 1.6 per cent.

Table. India – Land Use Categories (million ha)

Classification	1960-61	1970-71	1980-81	1990-91	1993-94
Reporting area	298.5	305.0	305.0	305.0	305.0
Area under non-agricultural uses	14.8	16.5	19.7	21.1	22.0
Barren and un-cultivable	35.9	28.2	20.0	19.4	19.0
Cultivable waste	19.2	17.5	16.7	15.0	14.5
Old fallow	11.2	8.8	9.9	9.7	9.7
Forests	54.1	63.9	67.5	67.8	68.4
Permanent pastures	14.0	13.3	12.0	11.8	11.2
Land under misc. trees etc.	4.5	4.3	3.6	3.8	3.7
Net sown area	133.2	140.3	140.0	143.0	142.1
Current fallow	11.6	11.1	14.8	13.7	14.3

The proportion of total area operated by marginal farmers increased from nine per cent in 1970-71 to nearly 15 per cent in 1990-91 while the proportion of large farmers declined from about 31 per cent to 17 per cent in the same period. The size of average holding is very unevenly distributed among the states. States with relatively large average size of operational holding are a mixed lot – they include states with large tracts of barren lands *e.g.* Rajasthan, Maharashtra and Madhya Pradesh on the one hand and agriculturally advanced state like Punjab on the other. These trends in farm size changes will have a profound effect on the future agricultural development strategies.

Table. Livestock and poultry population (million)

Category	1982	1987	1992
Cattle	192.45	199.69	204.53
Buffaloes	69.78	75.96	83.49
Sheep	48.76	45.70	50.79
Goats	95.25	110.20	115.28
Total (Livestock)	419.59	445.28	470.15
Total Poultry	207.74	275.32	307.10

Although India's population growth rate has slowed from 2.1 per cent in 1980s to 1.8 per cent in the 1990s and is expected to slow further in the coming decades, yet the population is projected to reach 1.33 billion by 2020 from the current one billion. The urban share of total population is projected to increase from 26 per cent to 35 per cent of the total population. Although incidence of poverty is falling, it is estimated that in 93-94 (upto which data is available) 320 million people constituting 36 per cent of the population were below the officially defined poverty line.

The nature of the poverty line has been shifting. About 30 years ago 48.4 per cent of those living in rural areas were poor and 20 per cent of those living in the urban areas were classed as poor. Recent studies show that the number of poor in urban areas have been increasing at relatively higher rate compared to the rural areas. At present those below the poverty line in rural sector constitute 37 per cent of the population while in the urban sector the percentage is 32 per cent. In the context of poverty alleviation, therefore, emphasis will be required to be placed both on production of food by the poor as well as on the availability of food for the urban poor.

It needs to be recognized that a large proportion of the rural poor are located in regions of low potential for food production *e.g.* arid and semi-arid areas, hilly regions, degraded land and forest areas. Widespread hunger and malnutrition are the direct manifestation of poverty and will call for increasing efforts to produce more food at affordable price.

Increasing population and economic growth are changing patterns of land use making potentially unsustainable demands on the country's natural resources.

Since early fifties the net area sown was expanded rapidly at first but at a diminishing rate since 1970 to reach approximately 142 million ha at present.

During 1950s and 1960s areas under agriculture expanded substantially as the fallows were reduced and cultivable wastes were put under the plough. The net area sown increased from 119 million ha in 1950-51 to 133 million ha by 1960-61 and further to 140 million ha by 1970-71. Fallow lands declined from 28 million ha in 1950-51 to 20 million ha by 1970-71. Cultivable wastelands declined from 23 to 17.5 million ha.

Land use intensity *i.e.* fraction of net sown area to total geographical area increase from 36 per cent in 1950-51 to 40.5 per cent in 1960-61 and 43 per cent by 1970-71 where it has since stabilized.

- Cropping intensity *i.e.* gross sown area as per cent of net sown area increased from 111 per cent in 1950-51 to 115 per cent in 1960-61, 118 per cent in 1970—71 and 130 per cent by mid 1990s.
- While the contribution of increased area in the growth of agriculture has declined over time, that of productivity has increased. The yield of all crops grew at 1.5 per cent per annum between early 1950s and mid 1960s. the pace accelerated to 1.7 per cent in the 1970s and then to 3 per cent per annum between early 1980s and mid 90s. Unlike the gains in area, which benefited non-foodgrains, the gains in productivity accrued mostly to foodgrains.
- India's forest resources have been dwindling. According to the 'State of Forest Report' (1997) the total forest cover of the country is estimated at 63.34 million ha *i.e.* 19.27 per cent of the geographic area of the country. Of these the dense forest (crown density more than forty per cent) and open forest (crown density 10 to 40 per cent) occupying about 11 and 8 per cent of the geographic area respectively and mangroves occupy 0.15 per cent of the geographic area. The country has lost about 5482 sq. km. of forest cover since the 1995 assessment. By any estimate the area under forest is far below the national policy goals and many areas nominally under forest are being used for non-forest purposes. Similarly 'uncultivated lands' such as permanent pastures, miscellaneous tree crops, cultivable wastes and fallow is subject to increasing competition from uses other than feeding livestock.
- The growth of livestock population is an important source of competition for land. The increase in number of major classes of livestock.
- The area sown to fodder crops is not recorded. Information available from other sources provide an estimate ranging from 4 to 5.5 per cent of the net sown area and suggest that the area under fodder crops will have to increase to 10 per cent or more to support increasing livestock based activity.

The pressure on India's land and water resources is seriously threatening native plant and animal diversity. India has uniquely rich and diverse genetic base. With increasing agriculture and economic development the genetic pool is declining. This decline, if unchecked and poorly managed can have unforeseen and adverse consequences for the sustainability of agriculture of the region.

AGRICULTURE IN THE CHANGING GLOBAL SCENARIO

Steady globalization of trade has profound implications for future agricultural development. The diversity of India's agro-ecological setting, high bio-diversity and relatively low cost of labour provide potential for agricultural competitiveness in a globalized economy. It is expected that with increasing globalization of markets over the years there will be demands for agricultural intensification. This will also be favoured because of greater backward and forward linkages between agriculture and food industry.

Therefore, increase in production and productivity are bound to be strategically important to economy. Intensification will not only favour alleviation of rural poverty but will also improve resource conservation particularly in the small farming sector where farmers can be encouraged to take up organized production of high value crops such as fruits, specialty vegetables, flowers medicinal and aromatic herbs etc. Stronger demands for crops of the small farmers' will not only improve incomes and welfare but will also make investments in technology and resource conservation more attractive. The General Agreement on Tariff and Trade (GATT) and liberalization of global trade is bound to have impact on future land use and production pattern. Understanding the local, national and international environment under which agricultural production is taking shape will be crucial in developing our own strategies.

EXTENSION STRATEGIES

Since early fifties a number of public by funded agricultural development programmes have been sponsored. These have included programmes like the National Extension Service (NES) Blocks in 1953, the Intensive Agricultural District Programme (IADP) in 1961-62, the Intensive Agricultural Area Programme (IAAP) 1964-65, the High Yielding Variety (HYV) programme 1966-67 and the Small and Marginal Farmers' Development Programmes (SMFDP) in 1969-70. Though these programmes had a perceptible impact the efforts did not get replicated over different areas and categories of farmers.

In mid seventies based on pilot level project in Rajasthan Canal and Chambal command area a 'Training and Visit' (T&V) system of extension was promoted in different states. Extension efforts of the Indian Council of Agricultural Research through its research Institutes and the State Agricultural University were largely limited to demonstration of new technologies through such programmes as National Demonstration Project, Operational Research Project, the Lab to Land Programme and the Krishi Vigyan Kendras.

However, there appears much to be desired in the way that extension programmes are conceived and implemented. At present extension programmes are implemented in largely a top-down fashion leaving little scope for localized planning and action. Farmers are almost passive receivers and their involvement in the process of technology generation and adoption is almost absent. Extension services, at present, are almost exclusively in the public sector domain and there is no effort or institutional support for other operators *e.g.* the NGOs, the corporate bodies etc.

Extension programmes sponsored by the government operate largely in isolation and there appears a strong need to view the extension programmes as an integral part of the research and development process.

The challenges facing agricultural development call for fundamental changes in our approach to technology transfer/extension programmes. Changes are necessary in the context of changing economic environment following policy adjustments in relation to privatization, deregulation and globalization calling for greater efficiency and effectiveness of the extension system. More importantly there is need for

- Greater emphasis on providing producers with knowledge and understanding needed to overcome the problems or to exploit opportunities of their own specific production systems. Correspondingly there will be a need to de-emphasize 'package of practices' or the blanket recommendations, top down approach followed thus far.
- Shift in the focus of public extension systems from promoting inputs use to one on sustainable management of resources and improvements in the production system as a whole.
- Closer interaction between farmers, extension scientists and production system researchers in diagnosing problems and identifying location specific recommendations emphasizing participation and education rather than being prescriptive.
- Widening the range of extension delivering agencies. While the publicly operated extension systems will continue to be important, there will appear a greater role for NGOs, farmers' associations and corporate sectors in particular situations. Role of commercial suppliers of seeds, agrochemicals, machinery, vaccines and medicines in providing advisories, as is already being done in a limited way, will need to be encouraged and factored into public system's own priorities.
- Wider and more creative use of mass media in tune with current developments in information technology to get information across to the farming community whose ability to overcome constrains at farm level will increasingly depend on access to reliable and up-to-date information.

TECHNOLOGICAL NEEDS AND FUTURE AGRICULTURE

It is apparent that the tasks of meeting the consumption needs of the projected population are going to be more difficult given the higher productivity base than in 1960s. There is also a growing realization that previous strategies of generating and promoting technologies have contributed to serious and widespread problems of environmental and natural resource degradation. This implies that in future the technologies that are developed and promoted must result not only in increased productivity level but also ensure that the quality of natural resource base is preserved and enhanced. In short, they lead to sustainable improvements in agricultural production.

Productivity gains during the 'Green Revolution' era were largely confined to relatively well endowed areas. Given the wide range of agroecological setting and producers, Indian agriculture is faced with a great diversity of needs, opportunities and prospects. Future growth needs to be more rapid, more widely distributed and better targeted. Responding to these challenges will call for more efficient and sustainable use of increasingly scarce land water and germplasm resources.

Technical solutions required to solve problems will be increasingly location-specific and matched to the huge agroecological/climatic diversity. Detailed indigenous knowledge and greater skills in blending modern and traditional technologies to enhance productive efficiency will be more than ever before, key to the farming success and sectoral growth. Most technological solutions will have to be generated and adapted locally to make them compatible with socio-economic conditions of farming community.

New technologies are needed to push the yield frontiers further, utilize inputs more efficiently and diversify to more sustainable and higher value cropping patterns. These are all knowledge intensive technologies that require both a strong research and extension system and skilled farmers but also a reinvigorated interface where the emphasis is on mutual exchange of information bringing advantages to all. At the same time potential of less favoured areas must be better exploited to meet the targets of growth and poverty alleviation.These challenges have profound implications for products of agricultural research. The way they are transferred to the farmers and indeed the way research is organized and conducted. One thing is, however, clear – the new generation of technologies will have to be much more site specific, based on high quality science and a heightened opportunity for end user participation in the identification of targets. These must be not only aimed at increasing farmers' technical knowledge and understanding of science based agriculture but also taking advantage of opportunities for full integration with indigenous knowledge. It will also need to take on the challenges of incorporating the socio-economic context and role of markets.

With the passage of time and accelerated by macro-economic reforms undertaken in recent years, the Institutional arrangements as well as the mode of functions of bodies responsible for providing technical underpinning to agricultural growth are proving increasingly inadequate. Changes are needed urgently to respond to new demands for agricultural technologies from several directions. Increasing pressure to maintain and enhance the integrity of degrading natural resources, changes in demands and opportunities arising from economic liberalization, unprecedented opportunities arising from advances in biotechnology, information revolution and most importantly the need and urgency to reach the poor and disadvantaged who have been by passed by the green revolution technologies.

Another important implication of increasing globalization relates to the need for greater attention to the quality of produce and products both for the domestic and the foreign markets. This would imply that production must be tuned to actual rapidly changing product demand. Such adaptation to global markets would require state of the art research, which can be achieved only by setting global standards of research, focus on well defined priorities and mechanisms which permit close interaction of farmers with researchers, the private sector and markets.

CHALLENGE OF INDIAN AGRICULTURE

There are several concerns about Indian agriculture that we need to address. Even in the green revolution crops, rice and wheat, the Indian average yield ranks around 50th in the world. A weak agro-based industry, the problems of weak marketing, storage, transportation, credit support, etc, haunt us. Only 0.3 per cent of agricultural GDP is spent on agricultural R &D. The population is growing at an alarming rate of 1.8 per cent per annum. It will be around 1.3 billion by 2020 and would require 50 per cent additional foodgrains. A total of 200 million people are still below the poverty line and have insufficient access to food. Declining resource base, degeneration of natural resources, including soil fertility, water environment, etc, are some other concerns.

Food production systems centre around a few states only. Out of 17 big states, only 7 are self-sufficient and surplus in foodgrains. There was a significant decline in per capita intake of calories and protein between 1975 and 1990.

Inspite of all these problems, there is plenty of good news. India harbours a vast diversity of basic natural resources. Tremendous bio-diversity of agriculturally important micro-organisms exists, which are not yet been harnessed. We have a variety of crops, including quality fruits, vegetables, foodgrains, and a number of plantation and commercial crops. Indian fisheries resource has spread over 8,100 km of coastline on the east and west, and provides access to marine resources and transportation.

Very large proportion of area of various food crops is under low productivity category. It can be increased with inputs of high class science. There is a scope to cultivate sizeable portion of waste lands through soil amendment and introduction of agroforestry. Several million hectares of saturated soil can be potentially exploited from rainfed lowlands of eastern India. The India Exclusive Economic Zone (EEZ) of the marine sector, of about 2 million sq.km., has high potential harvestable yields. Only about 20 per cent of the cultivated area is covered under tractor cultivation, great scope of making quantism quantum jump in production and productivity exists, changing consumption patterns offer scope for diversified agriculture. Trade liberalization provides opportunities to reach for outside markets, which were not accessible otherwise. Great gains are possible through agro-based industrialization and overall rural development.

The cultivators and farmers of India have a rich heritage of traditional agricultural wisdom on crop and livestock husbandry and fisheries; based on the principles of the conservation of natural resources and environment. India has an extensive network of extension channels for dissemination of technologies generated by research institutes. India has a strong manufacturing capability and a large network of about 20,000 manufacturers and nearly one million village artisans for indigeneous production of all types of agricultural machinery. Therefore, inspite of several disadvantages, we have a great chance to leap forward, if we enforce the right policy initiatives and implement them.

BIOTECHNOLOGY AND INDIA AGRICULTURE

We need to recongnise that, just as the food reequirements of today's population of nearly 6 billion people could not have been met by the technologies of the 1940's, we cannot assume that current practices will feed the population of 8 billion expected by 2020. New approaches are needed in addition to the continued improvement of existing methods of crop and animal husbandry and food processing.

The strategic integration of biotechnology tools into India agricultural systems can revolutionise Indian farming and usher in a new era. Genetic engineering is clearly the most revolutionary tool to impact agricultural research since the discovery of genetics by Mendel. Prior to genetic engineering, the exchange of DNA material was possible only between individual organisms of the same species. With the advent of genetic engineering in 1972, scientists have been able to identify specific genes associated with desirable traits in one organism. For example, a gene from bacteria, virus or animals may be transferred into plants to produce genetically modified plants having changed characteristics. This method therefore allows mixing of the genetic material from species that cannot otherwise breed naturally.

Genetic engineering can be vital for an agrarian country like India. It can help in minimising the crop damage through disease and pest resistant varieties, reducing the use of chemicals, enhancing stress tolerance in crop plants, thus permitting productive farming on unproductive lands, etc. One could even extend the growing season of crops and minimise losses due to environmental factors on the one hand and increase the shelf life of fruits and vegetables, on the other, thus minimising losses due to food spoilage. We can expand the market vista and improve food quality. Biotechnology can also produce plants that possess healthy fats and oils, possess increased nutritive value, and create a whole range of higher value feeds. Biotechnology has even the power of producing biodegradable plastics, edible vaccines, etc.

It is only through the blending of the 'gene revolution' with our experience in the 'geen revolution' that we can reach our goal of 'evergreen revolution' and also 'nutritional revolution' The advantage of the gene revolution is that it is relatively scale neutral, benefiting big and small farmers alike, It is also environment friendly. Thus, it can be of great help to the smallest farmer with limited resources in increasing farm productivity through the availability of improved but powerful seed. It can also reduce a farmer's dependency on chemical inputs such as pesticides and fertilizers.

Modern biotechnology offers unlimited opportunity for enhancing genetic potential of crops and other commodities, management of biotic and abiotic stresses, bio-remediation and organic recycling. India is one of the main centres of agricultural biodiversity and its gene richness can greatly complement the developments in modern biotechnology. Likewise, new developments in GIS, remote sensing, and crop modelling, provide new opportunities for integrated management of natural resources. The revolution in informatics provides opportunities for sharing latest information for research and planning in a highly organised and efficient manner. With the globalisation of economy, the opportunities for value addition and post harvest management are also immense. With clever blending of technologies, the India farmer can usher in new era of confidence and performance. Sir Francis Bacon had once said: *'It would be an unsound fancy to expect that things which have never yet been done, can be done except by methods which have never been tried'* In the new context, we have to review the new role of biotechnology. Also, it is in this context that we have to view the rapid advances that modern biotechnology is making, including the area of transgenics. Let us focus on some of the developments in this area, since considerable interest and controversy has been created around the world today in this exciting area.

MOVEMENT IN TRANSGENICS

The advent of genetic modification over the last two decades has enabled plant breeders to develop new varieties of crops at a faster rate than was possible using traditional methods, with huge potential for further beneficial development.

Tobacco was the first plant to be genetically transformed\, in 1983, with cereals beginning in 1990, but it is only recently that products such as GM soya have reached the market place in Europe, and, for a variety of reasons, given rise to concern and controversy.

The concern must be addressed seriously if society is to exploit the new technologies appropriately. We will address this issue rather extensively a little later. Whereas in India, we have been discussing and debating the issues, the rest of the world is galloping ahead.

Table: Ag Biotech Product Sales

$ Millions	Sales			Annual Growth Rate	
	1994	1997	2002	1994-97	1997-02
Transgenic seeds and plants	20	405	2,030	173%	38%
Animal growth horomes	80	225	405	31	12
Biopesticides	35	65	110	23	11
Other	107	180	340	19	14
Total U.S. Sales	$242	$875	$2,885	53%	27%

Development, field testing, and commercialisation of transgenic crops, is now recongnized by the international scientific community to be an essential strategy for food security. Fortyfive countries, including India, have conducted transgenic crop field trials up to 1997. The activity in the developing countries has been shown in Table below.

Table: Releases of Transgenic Plants in Developing Countries (by Species and Introduced Trait)

Species	Introduced Trait					
	Total Fied releases	Herbicide Resistance	Insect Resistance	Virus Resistance	Product quality	Others
Maize	46	29	16	1	3	5
Soya bean	27	25	—	1	1	—
Cotton	24	16	15	—	—	—
Tomato	19	—	2	1	16	—
Potato	13	—	1	6	—	6
Subtotal	129	70	34	9	20	11
Other species	30					
Total	159					

The emphasis has been mainly on traits for herbicide, insect, virus and fungal resistance, and on product quality with a limited number of genes. Out of a total 159 releases. Argentina leads the way with 43, followed by chile, Mexico, Puerto Rico and the Republic of South Africa, with 17-20 each. India is far behind with just a few releases. There are fears relating to transgenic plants in society. These need to be addressed, not by brushing them aside, but by careful scientific analysis and educational programmes. All biosafety protocols must be followed rigidly. Let us deal with this issue more exhaustively now.

PUBLIC CONCERNS

The potential of biotechnology as a method to enhance agriculture productivity in the future has been accepted globally. However, because of its revolutionary nature, risk and uncertainty may be created by the process of genetic engineering and by the resulting genetically modified products. This may result from the introduction of a new unrelated DNA sequence into a recipient organism. The introduced DNA may have some unexpected effects on the cellular processes of the recipient organism. In addition to this, introduction of antibiotic resistance genes, as selection markers, pose serious implications in public health especially in genetically modified plants that are directly used for food purposes.

Risks are also associated with the genetically modified plants that are released into the environment. The nature of interactions with other organisms of the natural ecosystems cannot be anticipated without proper scientific testing. For example, modified plants with enhanced resistance to pests or disease threaten to transfer resistance to the wild relatives, which will have implications for biodiversity and ecosystem integrity. These, and many more, doubts plague the minds of common people. Therefore, it is very clear that if the biotechnology potential has to be completely realized, it has to be done in an extremely responsible manner. At each step, proper testing and scientific data have to provided. Not only this, proper regulatory and policy mechanisms have to be put into place.

It is therefore, absolutely essential, that conscious efforts be made especially by the developing countries to initiate well-defined programmes for the development and regulations of genetically modified plants. To make this possible and to benefically, ethically and sustainably reap the benefits of biotechnology, there is a pressing need for scientists, researchers, policy makers, NGOS, progressive farmers, industrialists, and representatives of the government, to come together on a common forum in order to discuss the common concerns and find solutions to them.

Will modified plants transfer their introduced genes into wild relatives growing nearby? Will modified plants that produce new compounds disrupt the 'balance of nature' in some way?

Could the planting of a restricted number of cultivators leave crop plants more susceptible to disease? Could the planting of a restricted number of cultivators lead to a reduction in biodiversity (of crop plants, weed species, insects and microflora in the fields in question)?

Will genetically modified plants be able to avoid the factors that regulate natural populations and thereby change the usual ' balance' between populations? Similarly, there are concerns about the social and economic effects. For example, some of these concerns are:

- How will the structure of farming (particularly in developing countries) be affected by biotechnology?
- How will patent laws affect traditional breeders 'rights (*e.g.* the right to save seed from one year to the next)?

- Will plant breeding be left increasingly in hands of a few companies, and if so, what effects might this have?
- Will some countries be 'plundered' for their genetic resources?
- We also have a set of ethical and moral issues that need addressing. In particular,
- The consumer has the right. What will consumers be told about the new food products?
- Is it acceptable to 'interfere' with nature through genetic engineering?
- Do we have the right not to use all means available to improve crop plants, especially when so many people are under or malnourished?

Then there are several regulatory issues. Among these, the three key issues are:

- Do current regulations give sufficient protection to farmers, consumers, those who have invested in, and those engaged in research?
- Is there sufficient international legislation to ensure environmental protection?
- Do current regulations compromise the competitiveness of biotechnology companies by being excessively restrictive?

The Royal society of London had appointed a group of experts to examine various aspects including the scientific evidence concerning the risk of transfer of genes from Genetically Modified (GM) crop plants to wild species and non-GM crops, the uptake of genes from GM food by the digestive system, and the current state of the regulatory system.

The experts concluded that the chances of gene transfer happening are slight, provided the regulatory processes are followed, but that this must be kept under consideration.

The experts also concluded that the uptake of genes via the food chain is not a new issue because genes (*i.e.* DNA) are normal constituents of the human diet. Many products from GM Plants, such as sugar prepared from GM tomato paste, are so similar that they are regarded as 'substantially equivalent' others, for example flour from GM soya, may contain a new gene or its product, although many of the purification processes involved in food production will destroy and DNA present in the raw material.

Some GM foods have been produced using an antibiotic resistance 'market' gene, which is a laboratory device designed to identify genetically transformed plants.

The Royal Society has shown concern about the use of such genes in food products and it suggested that any further increase in the use of such markers in the human or animal food chain would be undersirable. Gm crop plants have been produced to improve insect tolerance and virus resistance, and to include herbicide tolerance, so that transferred to non-target species of plants, and that the development of resistance by target pests is minimised, the regulatory authorities must be assured that any negative effects would be no greater than those resulting from conventional procedures; and that any long term effects on the environment and ecology would be closely monitored, with statutory restrictions in place to control marketing; and also that best practice advice was adopted by growers.

The reliance on a case by case approach to the legislation, may result in lack of analysis of the overall impact of the technology on agriculture and the environment, and of the long term effects of gm. Although mechanisms are already in place to regulate many individual aspects of GM technology, there is no means for looking at GM technology as a whole. As independent, overarching, regulatory body is needed to span departmental responsibilities, monitor the enforcement of existing or future regulations, and strengthen the guidelines to growers of GM crops, such as those specifying the isolation distances between Gm and non-GM crops. The body should also review and monitor the membership of advisory committees and regulatory bodies.

BIOTECHNOLOGICAL APPLICATIONS RELEVANCE TO THE INDIAN FARMER

Increased food production must keep pace with the country's population increase. Not only the quantity of food made available should improve. To keep pace with the present population growth and consumption pattern, India's food requirement has been estimated to cross 225 million tons by 2005 AD. This would mean an annual agricultural growth rate of 6.7 per cent, a daunting task considering the rapidly shrinking resource-base and fast declining input-use efficiency in major cropping systems.

Not withstanding the impressive gain in the agricultural production, the vast agricultural potential still remains highly under-utilized. There are serious gaps both in farm yield realization and technology transfer, as the national average yield of most crop varieties of most crop varieties is low.

Not many superior varieties is mot pulses, many oilseeds and vegetable crops have been evolved in recent past. In case of vegetables the varieties which were imported or developed many decades ago are still being grown by the farmers. Their yields are low and their inherent ability to respond to improve production technology is also limited. Farmers have accepted bybrids in my vegetable crops such as cabbage, tomato and capsicum.

RESEARCH AND TECHNOLOGY TRANSFER

The Central and State governments continued to invest heavilyu in research, hence the Indian public sector research system has increased in size in real terms but by mid 90s levelled off. The number of hybrids and varieties developed and released by ICAR Crop Improvement Projects and State Agricultural Universities in major and minor crops are enormous and have been well documented. Private research to develop superior hybrids was initiated by few seed companies in mid 60s. By 1987-88, the number increased to over 12 and in 1997-98 it is estimated at 34. With the governments recognition of provate sector's performance capabilities and with "New Seed Policy" announced in 1988, the number of private companies involved in crop improvement programmes have been steadily increasing (Table below).

NUMBER OF PRIVATE COMPANIES ENGAGED IN CROP IMPROVEMENT PROGRAMME (1987-1998)

For want of financial, infrastructural and trained human resources, most private companies specialize in breeding superior hybrids in few crops characterized by low volume, high value and amenable to hybrid technology. In addition to the ICAR and Agriculture University research system, about 13 private sector seed companies are actively involved in vegetable R&D. Some of them have joint ventures with multinational or companies specializing in vegetable crop research and utilize research facilities available with their overseas partners.

Table: Number of Companies with Crop Improvement Programmes

Crop	1987-88	1993-94	1997-98
Pearl Millet	12	23	25
Sorghum	10	12	15
maize	06	15	20
Sunflower	10	30	20
Cotton	09	21	34
Hibrid Rice	-	4	6

Few have created facilities for field testing imported and public bred varieties. The role of private sector in vegetable research is primarily in the development of F hybrids and few open pollinated varieties.

Table: Hybrid Vegetable and Melon seeds of Private R&D available in the market (1996-97)

Crop	No. of Companies	No. of hybrid Varieties
Tomatoes	11	45
Brinjal	10	38
Chilli	8	27
Bhendi	13	25
Cabbage	11	40
Cauliflower	8	21
Ridgegourd	6	8
Watermelon	5	13
Muskmelon	2	7

Number of private/public hybrids in major crops in market during 1995 are compared with the number in 1997.

In pearl millet, sorghum, maize and cotton number of hybrids introduced by private, companies has been steadily increasing. Many hybrids especially in paddy and cotton have regional adaptability. The number of hybrids which are accepted by the farmers and may be grown on 2 per cent or more area is much lesser, so also the number evaluated in coordinated project trails or identified for release. Many hybrids recently supplied by the private sector embodies new technology which is either imported or developed by private companies. Private R&D's real investment in research has quadrupled between 1986 and 1997. Subsidiaries or joint ventures with multinational companies account for 30 per cent of all private seed industry research in 1997. Both domestic companies and multinationals companies have started research programmes on rice and mustard. Recently, Monsanto has also focused on hybrid wheat.

Table: Number of Private and Public Bred Hybrids in Major Crops in the Market during 1995 and 1997

Number of Hybrids in Market

	1995		1997		
Crop	**Private**	**Public**	**Private**	**Public**	**No of Private Hybrids grown on 2% plus area**
Pearl Millet	50	04	60	06	14
Sorghum	22	04	41	05	6
Maize	57	03	67	03	12
Sunflower	47	05	35	06	10
Cotton	73	15	150	15	19
Hybrid Rice	4	4	8	4	-

Few private sector companies have also initiated joint ventures (*e.g.* Mahyco-Monsanto Biotech, ProAgro-Agro-Evo) to transfer specific genes to their product lines to provide effective protection against pests. Some of them have taken initiative to treat their seed with high-tech chemicals to provide protection against sucking pests (aphids, jassids, thrips) during seedling and grand growth crop stages and diseases (downy mildew in bajra) during flowering.

COMMERCIAL SEED PRODUCTION

Majority of certified/commercial seed is produced under contract with trained seed multipliers in suitable agro-climatic conditions. Both public and private companies make contact with progressive farmers or trained seed multipliers and negotiate terms and conditions of production. Individual seed multipliers have small areas to contract and this greatly increases the cost of production and administration. To overcome this situation, contracts are arranged through local organizers, who in turn contract individual farmers, consolidate individual fields intoblocks and supply to the company the ultimate product.

Companies generallyprepare a technical package for seed multiplication specifying the kind of soil needed, isolation distances required, staggered plantion in case of a hybrid if needed, distances among rows and plants, depth of planting, time of nutrient and spray applications, frequency and time of irrigation, mode of harvest especially in case of hybrid seed etc. They prefer to contract between 300-400 acres in a village and designate such an area as a "seed village". They create seed processing facilities and place technically trained staff to guide seed multipliers, monitor actual production and supervise seed processing. Generally one such unit serves a cluster of 5-10 seed villages

Only seed lots meeting genetic Purity, Physical Purity, germination and disease standards prescribed by the government of India, are packed and commercially sold. User-farmers have become very conscious not only of genetic purity and germination, but also of physical attributes of seeds. They prefer bold, shining and plump seed, unaffected by mould The companies therefore, organize seed production in areas with appropriate agro-climatic conditions and no rains in the post-flowering period of seed crop. They also ensure proper cleaning to remove undersized, premature, broken seeds or inert matter on cleaner-cumgraders, gravity table, de-stoners and thus improve physical attributes of seed lots. Seeds are treated with recommended pesticides to prevent them against stored grain and soil pests. To add value to their seed, many companies now treat them with specific chemicals to provide protection agains sucking pests during seedling stages or specific diseases such as downys mildew for 40-45 days after emergence.

The seed production of most hybrid crops is generally undertaken in Western and Southern parts of the country because of congenial agro-climatic conditions and possibility of taking two seed crops in a year if necessary.

The prominent states are Gujarat, Maharashtra, Andhra Pradesh, Karnataka and Tamil Nadu. In contrast, the Northern states from Punjab to West Bengal are dominated by rice, wheat, and legume crops which are less profitable. Consequently there is a major difference in the cadre of seed industry between the North and West/South of India. Technology of seed production is moving at a rapid pace. Hence the companies have to ensure that they pass on the technology to the contract seed multipliers to harvest high yields of good quality seeds. Most public bred hybrids and variety seeds are certified while the proprietary hybrid seeds are sold under truthful label. All seeds meet minimum prescribed standards but for proprietary hybrids quality monitoring is through their in-house system.

PRIVATE HYBRID'S SHARE OF SEED MARKET

Private sector has played a major role in supplying quality seeds of both private and/or public hybrids. It is interesting to note that 69 per cent of the pearl millet seeds produced during 1996-97 season was of private hybrids. Other crops where private research plays key role in seed supply are sudangrass sorghym (100 per cent), sunflower (98 per cent), Maize (90 per cent), sorghum (52 per cent) and cotton (50 per cent). The contribution of private research in value terms is also steadily increasing. The share of research hybrids to total turnover in these crops was about 70 per cent in 1996-97 as compared with 46 per cent in 1990-91. Private hybrids are successfully meeting the regional needs of early maturity, tolerance to pests, unpredictable agro-climatic condition and soil variations.

GOVERNMENT INTERVENTION: LEGISLATION AND POLICY ISSUE

At the highest level, government policy is to reduce controls and to make the economic system more transparent in practice however, there is potential proliferation of legislation affecting seed sector. Given the current climate of economic development there is an urgent need for positive relationship between the government and the private sector. In order to fulfil farmer's needs and GOI food production targets set for the year 2005, the existing misapprehensions should disappear to ensure private sector investment in infrastructural improvement and expansion to increase quality seed production.

Presently Indian seed industry is governed by the following Acts:

- Seed Act 1966
- Seed Rules 1968
- Seed Control Order 1983
- Essential Commodities Act 1955
- Package Commodities order 1975
- Standrads of Weights and Measures Act 1976
- Consumer Protection Act 1986.

Export Regulations and Quarantine:

- Plants, Fruits and Seeds (Regulation of import into India) Order 1989
- Urban Land Ceiling Act 1976
- State Acts of land acquisition/land use
- State Acts for the control/movement of crops and seeds.

The policy framework hitherto has been regulatory with excessive legislation. Above plethora of legislation covering the seed sector is out of date and restrictive rather than progressive. The multiplicity of Acts, rules and administrative orders have resulted in over regulation of the nascent industry.

BIOTECHNOLOGY FOR SOCIETAL DEVELOPMENT IN INDIA

The application of biotechnology for societal development has always received special attention since it benefits socially and economically disadvantaged sections of the society. Through this programme a number of projects are implemented for economic and technical empowerment of the beneficiaries. Efforts are made through training, demonstration and appropriate R&D inputs for providing tailor made solutions for location specific problems of SC/ST, women and rural population using modern biology and biotechnology. The following pages provide the highlights of the achievements.

PROGRAMMES FOR SC AND ST POPULATION

Biotechnology-based Programme for SC/ST population has benefited around 15,000 people directly and about 45,000 indirectly through 54 projects during the year. The projects have been supported in the areas of cultivation of medicinal and aromatic plants, biofuels, production of biofertilizers and biopesticides, organic farming, aquaculture, mushroom cultivation as well as human health care interventions etc. The universities, public funded institutions, Krishi Vigyan Kendras and voluntary/non-governmental organizations have been involved in the implementation of the projects. The target groups have received the benefit of hands-on training and field demonstrations.

Vermicomposting: MR Morarka GDC Rural Research Foundation Jaipur implemented a project in Jalgaon district of Maharashtra, to benefit 2600 SC/ST target population on economic utilization of waste, agriculture residues and cow dung. Over 1780 MT of waste, agricultural residues and cow dung was utilized for vermicomposting through establishment of 28 community facility centres.

Beneficiaries are utilizing vermicompost on crops like onion, sugarcane, cotton, banana, papaya and wheat. Some farmers switched over from conventional crops to cash crops such as colorful capsicum, gerbera as well as inter cropping of safed musli, turmeric and ginger in banana plantation. Farmers have been trained to use bio-fertilizer, foliar sprays and mulching to use the agro-waste for fertility and management. Marketing linkages have been established with “Organic Shoppe”, through which farmers are getting 10-20 per cent premium for their organic produce. About 20 entrepreneurs have taken up vermiculture as commercial activity. extension Material in the local language was distributed to the farmers.

Mushroom cultivation: A project on promotion and production of essential oil, mushroom and vermicompost was implemented by RRL, Jorhat for revenue generation of the ST population of Arunachal Pradesh. Motivation and demonstrationcum-training programmes have been conducted in 13 villages of five districts on cultivation of citronella, production of mushroom and vermicompost. Demonstration units have been established at RRL Branch Naharlagun farm. Mushroom spawn, earthworms and citronella saplings were distributed to the beneficiaries. Several beneficiaries established vermicompost units on commercial basis and are selling vermicompost. The beneficiaries at Ziro, Jairampur, Doimukh, Manmao, Hetlong and Tengman have established oyster mushroom cultivation units. So far, 10 acres of land have been planted with citronella and 7 acres of land with patchouli at Sonajuli village of Papum Pare district.

Medicinal plants: A project on strengthening health care practices and conservation of medicinal and nutritious plants was continued at Herbal Folklore Research Centre, Tirupati for implementation in 100 villages of two Mandals in Andhra Pradesh. Training workshops were organized with the help of Yanadi tribal people for community health volunteers on identification and sustainable harvesting methods of 20 species of locally used medicinal and aromatic species. The reputed healers were encouraged to prepare medicines needed for PHC from the identified plants. About 6000 seedlings were distributed to the beneficiaries for planting in kitchen gardens. Green health kits (samples of herbal medicines) and educational material in local language containing identification, cultivation techniques and preparation of medicines with their therapeutic properties were distributed to the trainees.

A project on creating awareness and promotion of medicinal plants for the production and marketing was implemented by RRL, Jorhat to benefit ST population of Arunachal Pradesh. Awareness and training programmes were organized on the importance, scientific cultivation and marketing of Piper longum, Kaempferia galanga, Andrographis paniculata, Stevia rebaudiana, Rauvolfia serpentina and Mentha piperita. Three nursery-cumdemonstration centers were established at Roing (1000 ft msl), New Bomdila (8160 ft msl) and Ziro (4350 ft msl) and 19000 saplings were distributed to ST farmers/NGOs for introduction and propagation in their fields. At high altitude, villagers were advised to cultivate Aconitum heterophyllym, Coptis teeta and Picrorrhiza kurroa. Beneficiaries have stopped collecting wild Homalonema aromatica, an important but rare aromatic plant of Arunachal Pradesh, and have started its cultivation in their farms. Ten community nurseries have been established with the help of Vivekananda Trust, Roing at ten villages of Lower Dibang Valley District. Twenty farmers were trained on cultivation and post-harvest management of medicinal plants and marketing tie-up has been established. Around 1266 local ST people participated in the awareness programmes.

A project on organic cultivation of medicinal and aromatic plants for income generation and sustainable development in Uttarkashi District of Uttaranchal was undertaken through Forests Conservation Programme of World Wide Fund for Nature, New Delhi. Demonstration-cum-Resource Material Production Centre established for training on vermicomposting and cultivation of species.

Construction of poly house, vermi-composting units and planting of seeds of Rauvolfia serpentine completed. Awareness and motivation camps were organized for the farmers. Extension material brought out in the local language for cultivation of Rauvolfia serpentina, Coleus barbatus and Pelargonium graveolens were distributed to the farmers.

A project on developing skills and economically sound vermicompost units utilizing distillation waste of Mentha arvensis (menthol mint) was implemented for the benefit of SC/ST communities of Lucknow and Barabanki Districts of Uttar Pradesh. Around 200 farmers were trained on menthol mint nursery raising, transplanting, distillation and practices followed in vermicomposting. Demonstrations were carried out in farmers' field using high yielding variety 'Kushal'. Farmers trained on organic farming started vermicompost production using distillation and other agro-waste. The yield data obtained from the demonstrations showed the possibility of reduction of chemical fertilizers by 75 per cent without affecting the yield.

BIOFUELS

A project was continued on cultivation of Jatropha in arid lands of Namakkal District in Tamil Nadu for production of biofuel and other products for SC/ST community by Centre for Research in Social Sciences, Coimbatore. A high tech nursery was established and plantlets of Jatropha were raised. Training programmes were organized for 223 members on cultivation of Jatropha in 31 acres of land. About 30,000 plants were supplied for cultivation. Awareness camps, training and demonstrations were conducted on Jatropha cultivation, oil extraction and soap making. SHG groups were trained in nursery techniques. Women self help groups were trained in various activities of jatropha cultivation and soap making. Extension material bought out in the local language was distributed to the beneficiaries.

Biocontrol agents: A project was implemented on development of eco-preneurship among the coastal community of Pondicherry by MS Swaminathan Research Foundation, Chennai. Sixteen SHGs involving women and youths were formed in Ariyankuppam, Nonankuppam, Kakayanthope and Veerampattinam villages. Demonstrations and trainings were conducted to educate trainees on basic aspects of IPM, importance of biocontrol agents, production and marketing of Trichogramma, importance of human nutrition, role of edible mushrooms and production and marketing of oyster mushroom. Micro-enterprise was developed for mushroom and Trichogramma production. Technologies were disseminated on biopesticide (Trichogramma), oyster mushroom production and vermicomposting to recycle the spent waste of oyster mushroom. Extension material brought out in the local language was distributed to the villagers.

AQUACULTURE

A project was undertaken on enhancing livelihood of SC/ST people in Orissa and Tamil Nadu by MS Swaminathan Research Foundation, Chennai for introducing various aquaculture enterprises to 250 participants in backyard ornamental fish culture and carp seed production. Fingerlings produced were distributed to the entrepreneurs for stocking. Various aspects on ornamental fish breeding, carp nursery and gold fish culture were disseminated to the self help groups, which helped them in earning a revenue of ₹ 200/month/household and an income of about ₹ 3200/from carp nursery and ₹ 8500/from vegetables and flowers as bund crops. Community ponds were effectively utilized by stocking with rohu, catla, mirgali, Chinese and Indian carps and marketable size fish was harvested. Technologies transferred to the entrepreneurs on community pond aquaculture, carp nursery and ornamental fish culture and extension material in local language were distributed to the beneficiaries.

A project on utilization of integrated marine living resources by SC youth from Tuticorin region of Tamil Nadu was undertaken by College of Fisheries, Tuticorin. Lobster fattening and crab culture techniques were standardized and demonstrated to the beneficiaries, who have grown lobsters in cages for about 4 months and produced 250-350gms size for marketing. The beneficiaries were helped in availing credit facilities and bank loan. In the project around 120 people of target population have been trained. A manual on crab and lobster fattening technology was brought out for the distribution to the beneficiaries.

A project on freshwater prawn culture was implemented in village Naubasta Kalan in Allahabad District by University of Allahabad. Demonstrations were conducted on prawn hatchery technology and standardized methods for biocontrol of prawn disease for integrated giant freshwater prawn (Macrobrachium rosenbergii) culture. Trainings were conducted on scampi culture, breeding and seed production and microbial inputs were utilized for feed silage preparation using indigenous technology. Apiculture activity was introduced and disseminated to the target population and extension material was distributed to the beneficiaries.

GENETIC COUNSELING

Tribal population of northeast region was screened for anaemia and nutritional deficiencies through a programme implemented at Vivekananda Institute of Medical Sciences, Kolkata. The tribes of Arunachal Pradesh, Assam and Tripura displayed high incidence of anaemia. Awareness programme was organized for school children on inherited anaemia in pre-marital and pre-natal population. More than 1500 families from 80 villages were educated on anaemia, thalassaemia and hemoglobinopathies and counseling was provided to them. Leaflets were distributed for awareness of anaemia with special reference to Thalassaemia in local languages.

A project on molecular epideomiology of alcoholism was implemented jointly by Centre for Cellular and Molecular Biology, Hyderabad and Tribal Research Centre, Udagamandalam, Nilgiri District to study Kota tribe population in Nilgiri hills. General health awareness camps were arranged in seven villages for periodical medical check up. Blood samples were collected from 620 individuals and clinical data on health profile revealed that the disease morbidity is more for hypertension, heart related problems and diabetes. Genealogies indicated that the alcohol drinking habit is running in the families for the past two to three generations. Study undertaken on the health awareness camps and medical camps were found beneficial in counselling the tribal population.

FLORICULTURE

Over 180 women from Shillong, Itanagar and Chandigarh were trained in cultivation of commercially important and hybrid species of orchids. A total of 118 women of Shillong have raised 6 species *viz.* Dendrobium fimbriatum var. oculatum, D. longicornu, D.lituiflorum, C. pendulum, C. aloifolium and C. giganteum in their backyard. These plants have started blooming. The beneficiaries have started selling their harvest in local market. Through S&T Council, Bhopal 200 women were trained in cultivation of gladiolus, marigold, mogra, rose and tuberose.

FOOD PROCESSING

At College of Horticulture, Vellanikara, Thrissur, 88 women were trained in various aspects of entrepreneurship in preparation of cocoa products including molding, packaging and marketing. Twenty-two women have already set up their production unit and are selling their products in local market.

In another project at Fisheries College, Tuticorin, 500 women were trained in cook-chill technology for seafood and various parameters for quality assurance *viz.* estimation of moisture content, bacteria load, presence of bacterial species, the nutritive value, shelf-life etc. The trained women were grouped in 48 SHGs for establishing market for seafood products.

At the Central Food Technological Research Institute, Mysore data on biochemical and microbial analyses for nutritionally important factors like protein, amino acids, isoflavones, genistein, diadzein and methods for improvement of shelf life were collected on for 'Tofu' preparation. Also a byproduct of Tofu preparation, Okara, was found to be rich in protein and amino acids has been used as a food supplement. In another project at CFTRI, 17 women were trained in preparation of various food products of high nutritive value, methods of removal of anti-nutritional factors like phytic acid and trypsin inhibitor found in soya bean, better texture and improved shelf life. The food items included fermented products of soya bean and fruit pulp. Also 34 women were trained in spirulina production and its application as health food. Through CFTRI, technologies for rural bakery are being disseminated to rural women in three villages.

POULTRY AND LIVESTOCK FARMING

The 'Action Research on Health and Socio Economic Development', Bolangir, Orissa, identified 100 women and trained them in poultry rearing. The beneficiaries were provided with parent birds of Carigold (dual purpose) and Vanaraja varieties. The beneficiaries were able to earn ₹ 1000/per month through the sale of eggs and birds. The beneficiaries were also trained in feed formulation using local ingredients, management practices and entrepreneurship skill. The trained women formed SHGs and are approaching a bank for setting up their poultry farms. The market is established through state owned apex cooperative known as OPOLFED.

Through three projects supported at Tamil Nadu Veterinary & Agricultural Sciences University and its Extension Centres at Namakkal, Chennai and Tiruchirapally, over 1200 women were trained in poultry farming including feed formulation, farm management/ health care and preparation of various food items such as quail egg pickle and ready-to-eat food products etc.

Aquaculture: Under a project at the North Eastern Hill University, Shillong, techniques were standardized for induced breeding for common carp and 4,00,000 fish seeds were reared up-to fries for distribution to the beneficiaries. Selected 92 women were trained in fish farming techniques including water quality testing, pond maintenance, induced breeding, harvesting and feed formulation using slaughter house waste. They were provided with fish fries for rearing in their own pond.

Over 470 women from Bhopal, Shimla, Solan and Bhubaneshwar were trained in polyculture fish farming. Interested women were also provided advance training in composite carp culture, integrated fish farming with duck/ poultry, pond management and value added products including fish processing. These women are selling their products in local market and are earning ₹ 3000/ per harvest/family.

Considering the opportunities in ornamental fish, a project was supported at the University of North Bengal. The identified women, 57 from Darjeeling and 54 from Jalpaigudi district were trained in rearing and breeding of guppy (Poecilia reticulate) in cemented cisterns and angel (Pterophyllum scalaer) in ponds. The women are being trained in management skill and marketing.

ORGANIC FARMING/ VERMICOMPOST

At the Sri Sathya Sai Institute of Higher Learning, Anantapur Campus, Andhra Pradesh, neem varieties with higher level of azadirachtin were identified in 5 villages. Thirty trees were selected and 40 women from two villages were trained in fruit collection, drying and storage. Specialised training on maintenance of neem processing gadgets, processing of neem fruits for production of cake, oil or powder, maintenance of aza-rich plantlets, hardening, harvesting and application of neem based biopesticide was provided to five resource persons

At the University of Agricultural Sciences, GKVK Campus, Bangalore, five villages of Bangalore and Kolar were surveyed for the availability of organic waste. The waste materials were mostly sugarcane trash, coconut coir pith and vegetable wastes. Demonstration programmes for 20 to 25 farmwomen from each identified village were organized. Six interested women were given onfarm training. The technique of Trichoderma enriched vermicomposting was perfected and demonstrated at University campus.

At MSSRF, Chennai, phosphate-solubilising local strains were collected and screened. Simultaneously several awareness programmes on application of biofertilizer were organized for the villagers of Rediarchatram block. About 300 interested farmers were trained in large-scale production of biofertilizer.

Under the project support at Dangoria Charitable Trust Hyderabad, apart from training women in processing various items of food, 100 women from 13 villages were trained in production of vermicompost, neem-based biopesticide and their application in organic farming, About 100 vermicompost units were set up using the ring method. The beneficiaries produced over 130 quintals of compost. The beneficiaries were encouraged to raise nurseries of drumstick, papaya and creeper spinach at their backyard using vermicompost. The beneficiaries were able to sell excess vermicompost and planting material.

HORTICULTURE

Through the Tamil Nadu Agricultural University, Coimbatore, 54 progressive farm women were identified for training on cultivation of hybrid rice. The women were provided with quality seed their cultivation over an area of approximately four acre. The yield for the first harvest (December to May) was recorded at 2000 kg/ha. The beneficiaries were able to earn an additional income of ₹.20,000/per acre/year through three crops in an year. At Supi farms, Uttaranchal non-traditional vegetable crops were grown on farmers' land in a small way. An herbal garden was established. Incidence of pests to various crops was evaluated and farmers were trained on production of vermicompost and application of neem based biopesticide in horticultural crops.

MUSHROOM CULTIVATION

In an attempt to empower rural women through transfer of mushroom cultivation technology 93, women were trained at Himachal Pradesh Krishi Pradesh Vishwavidyalaya, Palampur and 750 at Tropical Botanic Garden and Research Institute, Kerala in mushroom cultivation. In addition, 15 women were trained in spawn production at TBGRI. The beneficiaries were given 15 days training in cultivation of Oyster and Agaricus species using wheat straw. The first harvest of about 1000 kg was sold @ ₹ 16/kg. The beneficiaries were also trained in various recipies of mushroom. Market linkage has been established for their products.

Under the project support at Dangoria Charitable Trust Hyderabad, apart from training women in processing various items of food, 100 women from 13 villages were trained in production of vermicompost, neem-based biopesticide and their application in organic farming, About 100 vermicompost units were set up using the ring method. The beneficiaries produced over 130 quintals of compost. The beneficiaries were encouraged to raise nurseries of drumstick, papaya and creeper spinach at their backyard using vermicompost. The beneficiaries were able to sell excess vermicompost and planting material.

SERICULTURE

A total of 550 women were trained for various activities including the organic cultivation of mulberry and rearing of cocoon and value added products from waste at the Central Sericultural Research and Training Institute, Mysore and at Vidyasagar University, Midnapore, West Bengal. The beneficiaries are getting an additional income through the sale of value added products and surplus vermicompost.

HEALTH CARE

Kidwai Memorial Institute of Oncology, Bangalore, Karnataka organized over 50 awareness camps on Carcinoma cervix to educate 2071 women for its prevention. Blood samples collected from 1830 women based on their medical history, 1812 were found negative and 18 positive cases were referred for treatment. A project on mental retardation including awareness generation among rual women, screening for early detection and genetic counseling was supported to Manovikas Bio-Medical Research. Over 7,000 families from three Gram Panchayats were surveyed. Ninety-one of these showing signs of mental retardation were shortlisted. The families were informed about the possible risk factors in future. Analysis of blood samples from these for genetic carrier showed only 4 cases of low level of fragile X chromosomes.

A project is supported at Saraswati Dental College, Lucknow for dental fluorosis. Water samples from 12 villages around Lucknow and Kanpur were analysed for fluoride content. Health camps were organized in 4 villages with a team of dental surgeon. A total of 2860 people were examined for dental fluorosis. The severely affected persons were advised for regular dental examination at SGPGI, Lucknow. They were also provided with calcium and cholicalciferol supplement. The blood samples from the severely affected boys and girls within the age group of 15 years and pregnant women were examined for their blood calcium level. Under another project supported at Mahavir Hospital And Research Centre, Hyderabad, Andhra Pradesh, blood lead levels in children and adults including men and women were measured to study environmental genotoxicity in women and children occupationally exposed to lead, a heavy metal.

Bibliography

A. B. Sharangi: *Seed Production of Selected Horticultural Crops*, Regency Publications, Delhi, 2014.

A. Fenwick Kelly: *Seed Production of Agricultural Crops*, Scientific Publication, Delhi, 2013.

Alexander C. Martin and William D. Barkley: *Seed Identification Manual*, Scientific Publication, Delhi, 2010.

Brajesh Tiwari: *Seed Production of Field Crops*, Oxford Book Company, Delhi, 2010.

Dhirendra Khare and Mohan S. Bhale: *Seed Technology*, Scientific Publication, Delhi, 2000.

Gurnam Singh: *Seed Pathology*, Gene Tech Books, Delhi, 2008.

Gurnam Singh: *Seed Quality*, Gene Tech Books, Delhi, 2008.

Gurnam Singh: *Seed Science and Technology*, Gene Tech Books, Delhi, 2008.

J.K. Sharma: *Seed Technology: A Practical Approach*, Westville Publishing House, Delhi, 2014.

K Ramamoorthy; K Sivasubramaniam and A Kannan: *Seed Legislation in India*, Agrobios Pubication, Delhi, 2006.

K. Raja, K. Sivasubramaniam and R. Geetha: *Seed Science and Technology : Questions and Answers*, Satish Serial Publishing House, Delhi, 2012.

K. Ramamoorthy and K. Sivasubramaniam: *Seed Technology Readyreckoner*, Agrobios Publication, Delhi, 2008.

K. Vanangamudi, G. Sasthri, S. Kalaivani, A. Selvakumari, Mallika Vanangamudi and P. Srimathi: *Seed Quality Enhancement : Principles and Practices*, Scientific Publication, Delhi, 2010.

K. Vanangamudi: *Seed Science and Technology : An Illustrated Text Book*, New India Publishing Agency, Delhi, 2014.

M. Prakash: *Seed Physiology of Crops*, Satish Serial Publication, Delhi, 2011.

O. P. Khedar, R. V. Singh, Y. K. Sinsinwar and Ved Prakash: *Seed Production Technology in Field Crops*, Pointer Publication, Delhi, 2013.

O.P. Verma: *Seed Production Techniques of Major Crops*, Daya Publication, Delhi, 2011.

P. Geetharani, V. Swaminathan and V. Ponnuswami: *Seed Technology of Horticultural Crops*, Narendra Publication, Delhi, 2012.

P.C. Trivedi: *Seed Technology and Quality Control*, Pointer Publication, Delhi, 2011.

Prabhakar Singh and B S Asati: *Seed Production Technology of Vegetables*, Daya Publication, Delhi, 2008.

Premjit Sharma: *Seed Legislation*, Gene Tech Books, Delhi, 2008.

S.P. Singh: *Seed Production of Commercial Vegetables*, Agrotech Publication, Delhi, 2001.

Sanjeev Agrawal, E.V. Divakara Sastry and R.K. Sharma: *Seed Spices : Production, Quality, Export*, Pointer Publication, Delhi, 2001.

Shalini Suri: *Seed Technology and Seed Pathology*, A P H Publication, Delhi, 2011.

Shyam Sundar Mondal, Mithun Saha and Kajal Sengupta: *Seed Production of Field Crops*, New India Publishing Agency, Delhi, 2009.

T.K. Chandra Bhanu and Vasudev Bhatnagar: *Seed Science and Technology*, Campus Books, Delhi, 2009.

Tribhuwan Singh and Kailash Agrawal: *Seed Technology and Seed Pathology*, Pointer Publishers, Delhi, 2001.

Uma Shankar Singh: *Seed Technology and Seed Pathology*, Anmol Publication, Delhi, 2008.

V.K. Agarwal: *Seed Health*, International Book, Delhi, 2006.

Y. S. Sreenivas: *Seed Production of Commercial Vegetables*, Oxford Book Company, Delhi, 2009.

Y. S. Sreenivas: *Seed Technology and Seed Pathology*, Oxford Book Company, Delhi, 2009 .

Index

A

Adequate 176, 184
Albizzia 162
Ambient 179, 182
Anaerobic 178, 191
Angiosperms 2, 4
Antimicrobials 179
Archegonium 118, 221
Aubergines 91, 100, 101

B

Beetroot 93
Bennettitales 223
Brassicas 95, 96, 140
Breathable 103
Broccoli 95, 96, 100
Brussels 95, 102, 105

C

C. Botulinum 178, 183, 185, 189, 191, 192
Cabbages 95, 96
Conjunction 177
Contamination 174, 181, 187, 191
Conventional 176
Cotyledons 113, 114, 115, 118, 120, 121, 143
Courgettes 97
Cucumbers 98, 99, 100, 102, 103

E

Eggplant 102, 103
Embryo 99, 112, 117, 118, 119, 122, 123, 124, 126, 127, 128, 131, 132, 133, 134, 145, 147, 148, 149, 150, 151, 152, 153, 154, 221, 224, 225, 226, 227
Endosperm 112, 113, 114, 115, 118, 120, 224, 225
Extrinsic 190

G

Gigantopterids 223, 224

H

Herbs 95
Hospital 284
Humidity 188

I

IBPGR 126
Implications 178
Inhibition 175, 181
Inoculum 186, 187, 188, 189, 190, 191, 192
Intrinsic 188, 190

Irradiation 180, 181, 185
ISTA 162, 168, 169, 172, 173

K

Kale 95, 100, 143, 144, 147

L

Lactobacilli 189
Lettuce 97, 100, 141, 142

M

Marrows 97
Megasporophyll 221
Melons 98, 101, 102
Microbial 174
Microflora 180, 182, 184, 189
Microorganisms 174, 175, 176, 177, 180, 181, 182, 183, 185, 188, 189, 190, 192
Mitochondrial 229, 230, 231, 232

O

Organoleptic 182
Oriental Brassicas 95, 96

P

Park 243, 246
Pasteurization 174, 179, 180, 181
Pathogenic 174, 175, 178, 180, 184, 186
Pelleting 106, 107, 108
Perisperm 114, 115, 118
Prediction 126, 127, 136
Progymnosperms 222, 224, 225
Pumpkins 97, 98, 102

R

Refrigeration 185

S

Samenanlage 224, 225
Science 247, 248, 277, 278
Sophisticated Method 107
Sowability 107
Squashes 97, 102
Symptoms 143, 144, 145, 148, 150, 151, 155

T

Technology 241, 245, 248, 249, 250, 278, 279, 280, 282, 284
Testa 114, 115, 118, 225, 227, 228
Turnips 95, 96, 102